Earth Observation Data Policy

Earth Observation Data Policy

Raymond Harris
Department of Geography, University College London

JOHN WILEY & SONS
Chichester · New York · Weinheim · Brisbane · Singapore · Toronto

Copyright © 1997 by John Wiley & Sons Ltd,
                    Baffins Lane, Chichester,
                    West Sussex PO19 1UD, England

                    National   01243 779777
                    International (+44) 1243 779777

All rights reserved. No part of this publication may be reproduced, stored in a retrieval system, or transmitted, in any form or by any means, electronic, mechanical, photocopying, recording, scanning or otherwise, except under the terms of the Copyright, Designs and Patents Act 1988 or under the terms of a licence issued by the Copyright Licensing Agency, 90 Tottenham Court Road, London, UK W1P 9HE, without the permission in writing of John Wiley and Sons Ltd., Baffins Land, Chichester, West Sussex, UK PO19 1UD.

Other Wiley Editorial Offices

John Wiley & Sons, Inc., 605 Third Avenue,
New York, NY 10158-0012, USA

WILEY – VCH Verlag GmbH, Pappelallee 3,
D-69469 Weinheim, Germany

Jacaranda Wiley Ltd, 33 Park Road, Milton,
Queensland 4064, Australia

John Wiley & Sons (Canada) Ltd, 22 Worcester Road,
Rexdale, Ontario M9W 1L1, Canada

John Wiley & Sons (Asia) Pte Ltd, 2 Clementi Loop #02-01,
Jin Xing Distripark, Singapore 129809

Library of Congress Cataloging-in-Publication Data
Harris, Ray, Dr.
    Earth observation data policy / Raymond Harris.
        p.   cm.
    Includes bibliographical references and index.
    ISBN 0-471-97188-X
    1. Remote sensing. I. Title.
    G70.4.H378 1997
    621.36'78—dc21   97-14901
                                                    CIP

British Library Cataloguing in Publication Data

A catalogue record for this book is available from the British Library

ISBN 0-471-97188-X

Typeset in 10/12pt Palatino from the author's disks by Vision Typesetting, Manchester
Printed and bound in Great Britain by Biddles Ltd, Guildford and King's Lynn
This book is printed on acid-free paper responsibly manufactured from sustainable forestation, for which at least two trees are planted for each one used for paper production.

Contents

Preface ix

Glossary of Acronyms and Terms xi

Acknowledgements xv

CHAPTER 1 Introduction 1
1 The importance of Earth observation data policy 1
2 Objectives of the book 4
3 Approaches to Earth observation data policy 6
4 Earth observation 8
5 Data policy 20

CHAPTER 2 Influencing Factors 21
1 Introduction and purpose 21
2 Objectives for Earth observation data policy 21
3 Growth of data supplies 26
4 Growth in formal user programmes 27
5 Economic and environmental issues 35
6 Commercialisation and market development 39

CHAPTER 3 Existing Policies and Policy Making Processes 43
1 Introduction 43
2 Global 43
3 Regional 56
4 National 61
5 Conclusion 65

CHAPTER 4 Physical Access to Earth Observation Data 67
1 Introduction 67
2 Direct acquisition 67
3 Programming satellite data acquisitions 72

4 Ground segments 75
5 Data distribution 85
6 Announcements of Opportunity 89

CHAPTER 5 Data Protection 91
1 Introduction 91
2 Legal protection of data 91
3 Encryption 97
4 Trends 98

CHAPTER 6 Data Pricing Policy 100
1 Introduction 100
2 Policy foundations 101
3 Pricing policy options 110
4 Conclusions 125

CHAPTER 7 Data Preservation 126
1 Introduction 126
2 Modes of working 128
3 Metadata 130
4 Policies and criteria 132
5 Conclusion 133

CHAPTER 8 Conclusions and Recommendations 134
1 Conclusions 134
2 Recommendations 140
3 Maximising value 143

Appendix: Members of the Committee on Earth Observation Satellites 144

References 146
Index 151

To Alexandra and Caroline

Preface

Earth observation is already a valuable technology for the collection of environmental information about our planet. Earth observation data are regularly used in numerical weather prediction, in the identification of agricultural fraud and in oil and gas prospecting. In addition, many organisations, including governments investing in space, see even greater potential for Earth observation in a wide range of scientific and monitoring programmes. The value of Earth observation has been shown in many projects around the world, so why is there a need for a book on Earth observation data policy? Why not just get on with doing the Earth observation itself?

There are three main motivations in writing this book. First, Earth observation data policy is as important as many technical issues in the development and maturity of the Earth observation sector: issues of access to data often have as much influence on the utility of Earth observation data as does the technology of Earth observation itself. Second, Earth observation data policy statements and debates are contained in many reports of working groups and government agencies, but unfortunately few achieve widespread circulation outside the originating group. This book provides a summary and analysis of the issues which many of these reports address. Greater clarity and explicit statements will contribute to the widening of the use of Earth observation data. Third, those organisations responsible for establishing Earth observation data policy often seem to lack good starting points: reinvention of the wheel springs to mind. This book provides those starting points and analyses of the issues.

This book is about Earth observation data policy. That is, it is about the policies which affect Earth observation data. What is it not about? Earth observation is used here as a synonym for satellite remote sensing, and the book is therefore about Earth observation data from space; it is not concerned with aerial photography or with data collected by other airborne platforms. Data from airborne platforms are typically controlled by the organisation which owns the instrument or the aircraft, and airborne data collection missions are restricted in space and time. Satellite Earth observation systems collect general-purpose data over large areas without explicit consent required from the areas observed. Therefore Earth observation provides a separate topic in data policy compared to airborne remotely sensed data. The book is also not about Geographic Information Systems (GIS) or about the policy issues of spatial data more broadly. The book will be useful to those concerned with GIS and with spatial data policy issues as an input from the Earth observation sector.

Glossary of Acronyms and Terms

AAAS	American Association for the Advancement of Science
AATSR	Advanced Along Track Scanning Radiometer
ADEOS	Japanese Advanced Earth Observation Satellite
ALMAZ	Russian radar satellite
ALOS	Japanese Advanced Land Observing Satellite
AO	Announcement of Opportunity
ASAR	Advanced Synthetic Aperture Radar
ATSR	Along Track Scanning Radiometer
AVHRR	Advanced Very High Resolution Radiometer
AVNIR	Advanced Visible and Near Infrared Radiometer
BNSC	British National Space Centre
CBERS	China/Brazil Earth Resources Satellite
CCD	Charged Coupled Device
CCSDS	Consultative Committee on Space Data Systems
CCT	Computer Compatible Tape
CD-ROM	Compact Disc – Read Only Memory
CEES	US Committee on Earth and Environmental Sciences
CEO	Centre for Earth Observation of the European Commission
CEOS	Committee on Earth Observation Satellites
CFC	Chlorofluorocarbon
CHEST	UK Combined Higher Education Software Team
CNES	Centre National d'Etudes Spatiales
CSA	Canadian Space Agency
DAAC	Distributed Active Archive Centre of EOSDIS
DALI	SPOT on-line quick look system
DMOP	Detailed Mission Operations Plan
DoE	UK Department of the Environment
DORIS	Doppler Orbitography and Radiopositioning Integrated by Satellite

Glossary of Acronyms and Terms

DRA	UK Defence Research Agency
EARSC	European Association of Remote Sensing Companies
EC	European Commission
ECMWF	European Centre for Medium Range Weather Forecasting
ECU	European Currency Unit
EM	Electromagnetic
Envisat	ESA environmental satellite
EOS	US Earth Observing System
EOSAT	US Earth observation satellite company
EOSDIS	US Earth Observing System Data and Information System
EPS	EUMETSAT Polar System
ERS	European Remote Sensing satellite
ESA	European Space Agency
ESF	European Science Foundation
ESOC	ESA European Space Operations Centre
ESRIN	ESA European Space Research Institute
EUMETSAT	European Organisation for the Exploitation of Meteorological Satellites
EWSE	European Wide Service Exchange of the CEO
FAO	UN Food and Agriculture Organisation
FD	Fast Delivery
FOIA	US Freedom of Information Act
FY	1. Chinese Earth observation satellite 2. Financial Year
GCOS	Global Climate Observing System
GDP	Gross Domestic Product
GEWEX	Global Energy and Water Cycle Experiment
GIS	Geographic Information System
GMS	Japanese geostationary meteorological satellite
GNP	Gross National Product
GOES	US geostationary meteorological satellite
GOME	Global Ozone Monitoring Experiment
GOMOS	Global Ozone Monitoring by Occultation of Stars
GOMS	Russian geostationary meteorological satellite
GOOS	Global Ocean Observing System
GTOS	Global Terrestrial Observing System
HDP	Human Dimensions of Global Change Programme
HRI	High Resolution Image of Meteosat data
HRV	High Resolution Visible instrument on SPOT
IACGEC	UK Inter Agency Committee for Global Environmental Change
ICSU	International Council of Scientific Unions
IEOS	International Earth Observing System

Glossary of Acronyms and Terms

IGBP	International Geosphere Biosphere Programme
IGOS	International Global Observing Strategy
INPE	Instituto Nacional de Pesquisas Espacias – Brazilian space agency
INSAT	Indian geostationary meteorological satellite
IOC	Intergovernmental Oceanographic Commission
IPCC	Intergovernmental Panel on Climate Change
IRS	Indian Remote Sensing satellite
ISIS	Intelligent Satellite data Information System
ISRO	Indian Space Research Organisation
JERS-1	Japanese Remote Sensing satellite
LACIE	US Large Area Crop Inventory Experiment
Landsat	US land resources satellite
LBR	Low bit rate
LRAC	Low rate Reference Archive Centre for Envisat-1 data
MARS	Monitoring Agriculture by Remote Sensing
MERIS	Medium Resolution Imaging Spectrometer
METEOR	Russian meteorological satellite
Meteosat	European geostationary meteorological satellite
Metop	ESA meteorological and climate satellite
MIPAS	Michelson Interferometric Passive Atmospheric Sounder
MOP	Meteosat Operational Programme
MOS	Japanese Marine Observing Satellite
MSG	Meteosat Second Generation
MTPE	US Mission to Planet Earth
NASA	National Aeronautics and Space Administration
NASDA	Japanese space agency
NERC	UK Natural Environment Research Council
NMS	National Meteorological Service
NOAA	US National Oceanic and Atmospheric Administration
Odin	Swedish Earth observation satellite
Okean	Russian oceanographic and hydrometeorological satellite
PAC	Processing and Archiving Centre for Envisat-1 data
PAF	Processing and Archiving Facility for ERS data
PDS	Payload Data System of Envisat-1
PDUS	Primary Data User Station for Meteosat data
PEP	Preferred Exploitation Plan
PI	Principal Investigator
PMB	UK Potato Marketing Board
POEM	Polar Orbiting Earth Observation Mission of ESA
PP	Pilot Project
RA	Radar Altimeter
Radarsat	Canadian Radar Satellite

Resource	Russian land resources satellite
SAR	Synthetic Aperture Radar
SCARAB	Scanner for Radiation Budget
SCIAMACHY	Scanning Imaging Absorption Spectrometer for Atmospheric Chartography
SDUS	Secondary Data User Station for Meteosat data
Seastar	US oceanographic satellite
SICH	Ukraine Earth observation satellite
SIR	Shuttle Imaging Radar
SPOT	Satellite Probatoire d'Observation de la Terre
SRIS	SPOT Satellite Reception des Images Spatiales
SWCC	Second World Climate Conference
TM	Thematic Mapper instrument on Landsat satellites
TOGA	Tropical Ocean and Global Atmosphere experiment
TOMS	Total Ozone Mapping Spectrometer
TOPEX/POSEIDON	France/US ocean topography mission
TRMM	Tropical Rainfall Mapping Mission
UN	United Nations
UNEP	United Nations Environment Programme
UNESCO	United Nations Educational, Scientific and Cultural Organisation
USGCRP	US Global Change Research Programme
WCRP	World Climate Research Programme
WDC	World Data Centre
WMO	World Meteorological Organisation
WOCE	World Ocean Circulation Experiment
WWW	World Wide Web

Acknowledgements

Since 1990 I have developed an understanding of Earth observation data policy through working with Logica UK Ltd, the European Commission, the European Space Agency and the British National Space Centre. This work has also brought me into direct contact with Earth observation organisations in Europe and North America.

In carrying out this work I acknowledge the valuable discussions with the following people on the subject of Earth observation data policy: Gerard Brachet, Stefano Bruzzi, Alan Cross, Derek Davis, Huw Hopkins, Roy Gibson, Roman Krawec, Robin Mansell, Livio Marelli, Cliff Nicholas, Pat Norris, Ichtiaque Rasool, Isi Saragossi, Lisa Shaffer, Bruce Smith, Zof Stott, Graham Thomas, John Townshend and Dave Williams.

I acknowledge the support of the Remote Sensing Unit of the University of London and the support of the Department of Geography, University College London.

1
Introduction

1 THE IMPORTANCE OF EARTH OBSERVATION DATA POLICY

1.1 Trade Secrets in the United States

Data from Earth observation satellites are valuable environmental information about the planet. The data are commonly regarded in a technical sense as exploitable information for scientific, operational or commercial applications.

There is, however, another dimension to the status of Earth observation data and this dimension is clearly illustrated by the US company EOSAT in its sale of Landsat Earth observation data. In the agreement which covers the purchase and protection of Landsat data from EOSAT there is the following warning which appears on the packages containing the data.

> These satellite data constitute a confidential trade secret of EOSAT. Use of these data by anyone other than the purchaser constitutes misappropriation of a trade secret. These data are proprietary information and have been disclosed in confidence to the purchaser and reproduction is prohibited. Reproduction of these data violates rights granted EOSAT by section 603 of the Land Remote Sensing Commercialization Act of 1984.

This package warning illustrates that Landsat data are not just humble technical information but are trade secrets. The data have potential value to the purchaser and they need protection. This protection is one part of the critical issue of Earth observation data policy.

Data policy is fundamentally affecting the development of the Earth observation sector. The conditions which govern access to Earth observation data, the distribution of the data, the preservation of the data and the price of Earth observation data are now vital to the exploitation of this important environmental data resource.

1.2 Assets in Canada

Across the border in Canada, the Canadian Space Agency (CSA) has explicitly built into the Radarsat data licence and distribution agreement the affirmation that (Radarsat 1996): 'The Licensee acknowledges that the Data is a valuable and unique asset and is disclosed to the Licensee on the basis that it represents confidential information.'

The agreement to use the Radarsat data identifies:

- the Canadian Space Agency as the owner of the data and the owner of the intellectual property rights to the data,
- Radarsat International Inc. as the Master Licensee, and
- the buyer, who buys a licence to use a copy of the data and does not buy the data set itself.

This shows that the Radarsat Earth observation data are valuable assets which implicitly have an intended purpose that is of value to the user, and indeed that they are confidential information. The control of the data is explicitly defined so that the user who buys a licence to use the data is strictly prohibited from distributing, leasing, selling or otherwise disposing of the data (Radarsat 1996).

1.3 Importance in Europe

In 1992 the European Commission established a senior, independent working group to examine the issues of Earth observation data policy for Europe. The working group was chaired by Roy Gibson, a former Director General of the European Space Agency and of the British National Space Centre, who was accompanied in the working group by 39 other Earth observation experts from around Europe.

In the foreword to their report the working group constructs the following argument (European Commission 1992a), which I quote at length because of its importance to the context of this book.

A major challenge facing mankind is to achieve a stable transition to sustainable development of the Earth's resources in the coming century. Understanding, monitoring and managing the Earth's environment has become a central issue in achieving this stable transition.

Earth observation by satellite is rapidly becoming an essential tool in the management of the Earth's resources, and for the study and monitoring of its environment and climate. Space-derived information is also of increasing value for the implementation of public policy in many areas. When responding to environmental challenges, there is a need to be proactive and encourage a greater use of Earth observation data.

While the potential and the importance of Earth observation to contribute to the understanding and management of the Earth's resources are very high, there are at present potentially incompatible or conflicting policies regarding the management, supply and exchange of data. A data policy acceptable to both the suppliers and to the wide range of users of data and derived information is required for the sector to mature and to meet the needs of understanding, monitoring and managing of the Earth's resources. It must also be a policy which encourages the widest possible use of Earth observation data and the identification and development of new uses of the data.

This quotation from the European Commission working group report shows how Earth observation data policy links to larger issues of scientific understanding and sustainable development, and at the same time demonstrates the limitations of conflicting data policies on the ways in which Earth observation data can be fruitfully used.

In June 1994 in Svalbard, the European Space Agency (ESA) presented to its Programme Board for Earth Observation a set of perspectives for a strategy for Earth observation from space. These perspectives were primarily geared to future programmes and missions, but the ESA presentation was clear in noting that data policy is an essential part of an Earth observation strategy. ESA further noted that any data policy should be aimed at ensuring the sustainable long-term availability of Earth observation data and derived services.

European industry was invited in November 1996 to present to ESA its views on the data policy for the forthcoming Envisat-1 mission. In their presentations, the European Earth observation industry was particularly concerned that ESA constructs a data policy for Envisat-1 which encourages the development and growth of the Earth observation sector and particularly fosters a wide use of Envisat-1 data.

1.4 Data Wars

On 16 February 1996 the *Times Higher Education Supplement* (THES) carried the following headline to lead an article on data policy: 'Data wars dispute hits research'. The THES article referred to the grave concern expressed at the 1996 meeting of the American Association for the Advancement of Science (AAAS) over access to international Earth observation data. The focus of the concern was on meteorological data obtained from weather satellites and the concern that US scientists and operational users have over access to European weather satellite data. Elbert Friday, director of the US national weather service, was quoted by the THES as saying:

There is a conflict. The private sector, which has readily available data from the US and other countries which carry out free and unrestricted policies, is in conflict with the public sector in other countries which, as part of their mandate from their governments, must obtain some of their costs by sales.

There were complaints at the AAAS meeting that the European Organisation for the Exploitation of Meteorological Satellites (EUMETSAT) was encrypting most of its geostationary weather satellite data and so restricting access to this environmental data resource. Researchers were especially concerned that the long-term climate data record will be damaged and sources of data for global change research will be diminished.

1.5 International Cooperation

International cooperation is fundamental to Earth observation (Abiodun 1993). The investment levels are sufficiently large so as to make international cooperation the only route for many nations to gain access to space. In a review of the structures for international cooperation in remote sensing for global change research, Thomas, Lester and Sadeh (1995) conclude that of all the challenges which face Earth observation, 'First and foremost, a coherent and standardized approach to data access, pricing and funding policies must be created for *all* remote sensing data relevant to global change research regardless of its type or source.'

The view of Thomas, Lester and Sadeh (1995) calls for a greater focus on agreeing data policy in Earth observation, a view that is echoed by the International Geosphere Biosphere Programme (IGBP) when Townshend (1996) notes, 'Data policy has become an increasingly important topic for the IGBP as more of its Core Project and Framework Activities move from planning to implementation stages.'

1.6 Chapter Structure

Earth observation is therefore not only about the capture and application of data about the planet Earth from space. It is also about the conditions of access to those data and therefore about data policy questions.

This book is concerned with Earth observation data policy, and as part of developing the context for the book this chapter continues by reviewing the objectives of the book, examining the sectors involved in Earth observation, briefly stating the physical basis for Earth observation and reviewing the main satellite Earth observation missions. The chapter is therefore concerned with the background issues which take us to the starting point of examining Earth observation data policy.

2 OBJECTIVES OF THE BOOK

Against this background of the importance of data policy to many organisations, this book examines the issues in satellite Earth observation data policy and provides a contribution to the current and future development of Earth observation data policies.

While there are data policies developed by the suppliers of data for each of their instruments or missions, and there are reports of various government and non-governmental agencies on the subject of Earth observation data policy, there is no single and comprehensive treatment of the subject in book form. This book draws together the main issues of a complex subject and presents the interrelated arguments that are affecting Earth observation data

Introduction

policy. The book has the following objectives:

- to summarise the Earth observation data policy positions of the main organisations in the Earth observation sector. Inevitably, the views of the data suppliers dominate because they own the data and set the policies for their data;
- to analyse the key Earth observation data policy issues, particularly access to data, data protection, data pricing and data archiving;
- to present the many and varied factors that are influencing the development of Earth observation data policy;
- to identify the recurrent themes that are likely to be present in Earth observation data policy for the next 20 years, and to recommend approaches to the questions of Earth observation data policy in ways which reduce the potential for conflict.

To reach these objectives the book is organised into eight chapters. Chapter 1 (this chapter) provides an introduction to the subject, including the importance of data policy, the physical basis of Earth observation and the main satellite Earth observation programmes. Chapter 2 describes the factors that influence Earth observation data policy, namely the objectives, the growth in data supplies and the growth of user programmes, the economic and environmental issues and the commercial context of the development of Earth observation data policies. Chapter 3 summarises the existing policies and policy making processes. The chapter presents international, regional and national positions on Earth observation data policy.

Chapter 4 describes the physical access to Earth observation data by direct acquisition and control of acquisitions, by special conditions of access and by the conditions that affect distribution. Chapter 5 is concerned with the data policy issues that flow from the protection of Earth observation data. Chapter 6 concentrates on what is a central theme of Earth observation data policy, namely pricing policy. Different pricing policy models are presented with arguments in favour and arguments against each model. Chapter 7 summarises the need for a greater concern for data preservation or data archiving, and Chapter 8 presents the conclusions of the book, and recommendations for Earth observation data policy development in the future.

Before examining the Earth observation data policy issues in detail, this introductory chapter describes the technical part of Earth observation, that is the physical basis of Earth observation and the satellite missions. The reader who is familiar with Earth observation technology may wish to scan this material quickly and move on to Chapter 2.

3 APPROACHES TO EARTH OBSERVATION DATA POLICY

3.1 Sectors Involved in Earth Observation

While there have been several categorisations of the sectors of users in Earth observation, a common approach is to employ three groups: scientific research, operational applications and the commercial sector. For each of these there are different issues in Earth observation data policy, and different perspectives on the importance of Earth observation data policy.

Scientific research that uses Earth observation data is concerned with the discovery of facts and principles about planet Earth, and the understanding of the processes in the geophysical, biospheric and atmospheric systems of the planet. This is, for example, the motivation of the US Earth Observing System (EOS) as the major part of Mission to Planet Earth (MTPE). The role of Earth observation data is particularly useful in the improved understanding of global change.

Operational users who use Earth observation data are those organisations which provide regular information products and services. They are typically funded by the public and produce information for the public benefit, e.g. the national meteorological services. Although such operational organisations are typically funded by the public sector, this is not always the case. These organisations can and do buy an operational service from commercial agencies and are increasingly seeking a proportion of their revenue from selling products and services.

The commercial sector in Earth observation comprises those organisations which build the space and ground infrastructure for Earth observation systems and those which use the data to provide value added products and services for end-user clients. The commercial sector is driven by growth, market sustainability and by making a profit in both the short and the long term.

3.2 Key Issues

These three sectors have an interest in Earth observation and Earth observation data policy. What are the key Earth observation data policy issues about which they have common concerns?

A recurrent theme is *stability*. For scientific research, and particularly for global change research, stability is essential to provide data from the same instrument under the same conditions for many years. For operational organisations their *modus operandi* is stability of products and services to their end-users, and so stability of data and data supply conditions is a core concern. In meteorology this has led to meteorological services being responsible for their own Earth observation missions, as is the case with the European national meteorological services and EUMETSAT. For the commercial sector, stability is essential so that investment decisions can be made with a correct understanding of the conditions of the future marketplace.

Simplicity is a second key issue in Earth observation data policy. As this book will show, the conditions of access to Earth observation data are probably subject to as many interpretations as there are people and organisations with informed opinions about the subject. Simple data policies which avoid the pitfalls of becoming too deeply entrenched in implementation are necessary for all three sectors in Earth observation.

Given that much of Earth observation is publicly funded, there is a concern for *fair treatment* to be applied and to be seen to be applied. The three sectors want fair treatment for themselves and for fair treatment to be seen to be applied to others. This means explicit conditions of access which do not arbitrarily favour one group or penalise another group.

Growth in the volume of data, in the types of information and in the number of users are all important factors for the research, operational and commercial sectors. The research sector wants access to data types from new instruments: atmospheric chemistry is an example of growth in the types of new space instrumentation. Operational organisations want growth in the number and range of users so that they can justify investment programmes. Commercial users want growth in the diversity of data and the user base so that they can increase their market size by offering new systems, products and services. Developing the right Earth observation data policies is critical to the growth that the three sectors are seeking.

There is widespread interest in *maximising the use* of Earth observation data. Most Earth observation data are produced as a result of public investment, and both the owners of the systems and the users of the data see benefits for all in maximising the use of these important environmental data in applications products and services.

A final common issue is the *sustainability* of the sector. A combination of high investment costs plus a high potential value of the data in the long term means that those involved in the sector see the value of a sustainable Earth observation sector that does not disappear shortly after applications have been brought to a mature stage. There is a common concern in Earth observation about the future patterns of investment and the different roles of the organisations involved in funding.

4 EARTH OBSERVATION

4.1 Sources of Information

This section provides a brief summary of Earth observation technology. For the purposes of this book, Earth observation is concerned with Earth observation from space satellites and does not include aerial photography or Earth observation performed by other instruments carried on board aircraft platforms. Airborne data collection is not considered because airborne surveys have restricted conditions of access. The data are collected for specific projects and have their own set of data policy conditions. Satellite Earth observation data are typically collected as general-purpose information with a wide variety of applications. This leads to differences in the conditions of access to airborne and satellite data.

Readers who want a fuller description of Earth observation and remote sensing technology may wish to consult one of the following texts: Campbell (1996), Cracknell and Hayes (1991), Curran (1985), Harries (1994), Harris (1987), Lillesand and Kiefer (1994), Mather (1987) and Williams (1995).

Two relevant journals which publish a wide range of technology and applications papers in Earth observation are the *International Journal of Remote Sensing*, published by Taylor and Francis Ltd, and *Remote Sensing of Environment*, published by Elsevier.

4.2 Physical Basis

Plane Waves

The physical basis of Earth observation is the measurement of electromagnetic (EM) radiation. EM radiation travels in the form of waves and these waves interact with objects such as the land surface or the atmosphere. The behaviour of the EM radiation can be used to give us information about the objects themselves.

Plane waves are the forms of EM wave energy in free space: they are characterised as having a constant phase over a plane perpendicular to the direction in which the wave is travelling. Plane waves are described by their wavelength (λ), amplitude (A) and phase (L), and these characteristics are shown in Figure 1.1.

Plane waves all travel at the velocity of light c, which is approximately 300 million metres per second. They have wavelengths which range from the gamma rays (3×10^{-9} m) through visible light (0.4–0.7×10^{-6} m), to very long radio waves of 3×10^6 m wavelength. Because their velocity is constant at the speed of light then the frequency of plane waves is inversely proportional to their wavelength.

Introduction

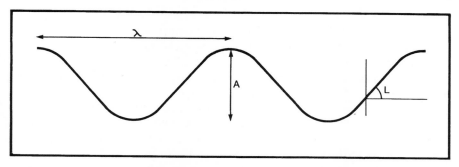

Figure 1.1 Plane Waves with Wavelength (λ), Amplitude (A) and Phase (L). Source: Harris (1987) published by Routledge

Wavelength and frequency are the main characteristics of plane waves which are used to describe EM energy in Earth observation. Earth observation in the visible and infrared parts of the EM spectrum commonly uses wavelength as the main descriptor, while Earth observation in the microwave region uses both wavelength and frequency. Table 1.1 gives the wavelengths and, where appropriate, the frequencies commonly used in Earth observation.

Blackbody Radiation

All bodies with temperatures above zero kelvin (0 K) emit EM radiation. The temperature of the object determines the wavelength of maximum EM energy emission by Wien's displacement law:

$$\lambda_{max} = \frac{a}{T}\,\mu m \tag{1}$$

Table 1.1 Wavelengths and Frequencies used in Earth Observation

Type	Wavelength	Frequency	Example
Visible	0.4–0.7 μm	—	SPOT HRV
Near infrared	0.7–1.5 μm	—	ADEOS AVNIR
Middle infrared	1.5–4.0 μm	—	Meteosat
Thermal infrared	8.5–12.5 μm	—	ERS-1 ATSR
Microwave	1.0–30 cm	1–12.5 GHz	—
X-band radar	2.4–3.8 cm	8–12.5 GHz	SIR-C/X-SAR
C-band radar	3.8–7.5 cm	4–8 GHz	ERS-1 SAR
L-band radar	15.0–30 cm	1–2 GHz	JERS-1 SAR

Note: 1 μm = 10^{-6} m.

where λ_{max} is the wavelength (μm) of maximum energy emission, T is the temperature of the object (K) and a is a constant (2898 μm K).

The hotter the body, the shorter the wavelength of maximum energy emission. For example, the λ_{max} of the planet Earth with a surface temperature of 288 K is approximately 10 μm, while the λ_{max} of the Sun with a surface temperature of 6000 K is approximately 0.48 μm.

The Earth and the Sun are blackbody radiators. A blackbody is not necessarily black, but is an object whose emissivity (ε) is 1; that is, it emits all the energy it absorbs. Planck's radiation law defines the relationship between energy, temperature and wavelength. It can be written for all wavelengths to give the total power (E) radiated per unit area of a blackbody in W m^{-2} as:

$$E = \frac{2\pi^5 k^4}{15 c^2 h^3} T^4 \tag{2}$$

where k is Boltzmann's constant (1.38×10^{-23} J K^{-1}); c is the velocity of light (3×10^8 m s^{-1}); h is Planck's constant (6.626×10^{-34} J s); and T is the absolute temperature of the blackbody (K). This relationship can be used to show that hotter bodies emit more energy at shorter wavelength than do cooler bodies.

Equation (2) can be simplified because the first term on the right-hand side of the equation consists entirely of constants, and the equation should also include an emissivity term (ε):

$$E = \sigma T^4 \varepsilon \tag{3}$$

where σ is the Stefan–Boltzmann constant (5.67×10^{-8} W m^{-2} K^{-4}).

A measurement of E will therefore give a measurement of a combination of T and ε. When a blackbody is being measured then ε = 1 and so its effect in equation (3) can be disregarded, but in some cases (e.g. cirrus clouds) ε is less than 1 so its effects must be estimated before T can be calculated correctly.

Absorption by the Atmosphere

The Earth's atmosphere has a powerful effect on the radiation passing through it. The Earth's atmosphere absorbs radiation over a wide range of wavelengths because of the large number of gases present in the atmosphere. These gases have important absorbing qualities:

- *Oxygen and ozone.* Energy of less than 0.1 μm is absorbed in the upper atmosphere by oxygen molecules and free oxygen atoms. Energy of 0.1–0.36 μm wavelength is absorbed by ozone.
- *Carbon dioxide.* There is a strong absorption band at 15 μm, and weaker bands at 2.5 μm and 4.5 μm.
- *Water vapour.* There is a strong water vapour absorption band at 6 μm and some absorption between 0.6 μm and 2 μm and at 3 μm.

Figure 1.2 shows the absorption of EM radiation by the atmosphere in the

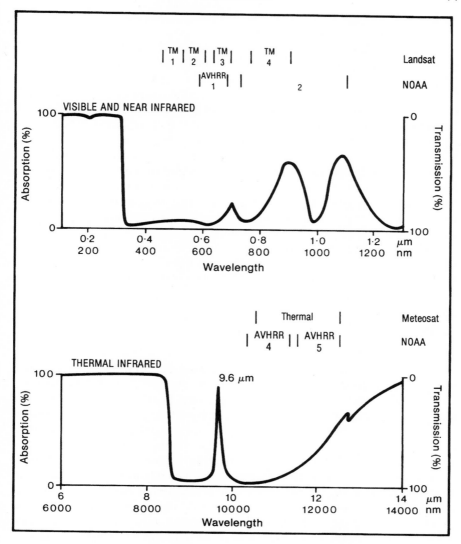

Figure 1.2 Absorption by the Atmosphere in the Visible, Near Infrared and Thermal Infrared Parts of the Electromagnetic Spectrum. Example Channels for Landsat, NOAA and Meteosat Satellites are Shown. Source: Harris (1987) published by Routledge

visible and near infrared and in the thermal infrared parts of the EM spectrum. Earth observation either uses the areas of low atmospheric absorption, the so-called atmospheric windows, to measure the surface, or it uses the regions of high atmospheric absorption to measure the characteristics of the absorbing gases at those wavelengths. For example, Figure 1.2 shows the wavelength ranges of example channels from the Landsat, NOAA and Meteosat satellite instruments: these channels are designed to concentrate on the areas of low atmospheric absorption so that the cloud, land, ocean or ice surfaces can be measured with little interference from the atmosphere.

By contrast, the Meteosat satellites also carry an instrument that measures radiation in the 5.7–7.1 µm channel, which is a part of the EM spectrum with high absorption by atmospheric water vapour. Measurement in this part of the EM spectrum gives information on the concentration of water vapour in the atmosphere and the Meteosat images in this channel show the patterns of water vapour distribution in the atmosphere.

Instruments to measure atmospheric chemistry typically use areas of the EM spectrum with high atmospheric absorption, while instruments to measure land, ice and ocean surfaces use areas of the EM spectrum with low atmospheric absorption.

Resolution

A further key characteristic of Earth observation instruments is the resolution of the instrument. Resolution can be described in three senses: spatial, spectral and temporal. Spatial resolution is commonly quoted as the pixel size. For SPOT this is 10 m for the panchromatic band, for Landsat Thematic Mapper (TM) it is 30 m, and for ERS-1 Along Track Scanning Radiometer (ATSR) it is 1 km. The spatial resolution is influenced by factors other than the pixel size, so the effective resolution is not the same as the pixel size although the pixel size gives a good first approximation of the spatial resolution of the sensor. The spatial resolution is affected by the spatial contrast in an image, the radiometric resolution of the instrument and the clarity of the atmosphere (Forshaw *et al* 1983). For some instruments (e.g. radar), the spatial resolution depends upon the processing after the data have been received. ERS-1 radar data can be presented in image form with pixels of 6, 12, 24 and 30 m: as the spatial resolution or pixel size increases then the level of speckle and noise in the data decreases.

Spectral resolution is the width of the waveband used in the Earth observation instrument. The early Earth observation instruments had relatively wide wavebands, e.g. SPOT channel 1 at 0.50–0.59 µm. New instruments are now being developed with much narrower wavebands so that the instrument can concentrate its measurements on certain chemical or physical characteristics of the surface being sensed. This is the case for the Medium Resolution

Imaging System (MERIS) (Bezy *et al* 1996) instrument on Envisat-1 where there are 15 wavebands which can have their widths programmed to as narrow a range as 0.0025 µm.

Temporal resolution is the period between repeat overpasses of the Earth observation satellite. Table 1.2 gives some example repeat or revisit times.

There is a trade-off between spatial resolution and temporal resolution. Earth observation satellites with small pixels have relatively narrow swath widths and capture a very large amount of data. Earth observation satellites with large pixels can have wider swaths and capture much smaller quantities of data. The ability of Earth observation satellites to transmit their data to the ground is a further constraint. The current maximum capability is approximately 150 Mbit s^{-1}. The downlink capacity for ERS Synthetic Aperture Radar (SAR) is 100 Mbit s^{-1}, and for Landsat it is 85 Mbit s^{-1}.

The instruments on the SPOT satellites are pointable up to 27° either side of the sub-satellite point. This capability allows them to achieve a revisit frequency as high as once per day, and typically once every 2.5 days.

Orbits

For most Earth observation missions there is a choice between two types of orbit around the planet Earth, the geostationary orbit or the polar orbit. Example orbits are shown in Figure 1.3.

The geostationary orbit is a position 35 900 km above the equator at which a satellite can rotate at the same speed as the Earth beneath. This makes the satellite appear to be stationary in relation to the Earth, and hence the term 'geostationary'. The global geostationary meteorological satellites and most of the telecommunications satellites all occupy this orbit. The images taken from geostationary orbit are images of the disc of the Earth below the satellite up to approximately 55° away from the sub-satellite point, and are consequently termed 'full disc' images.

Most Earth observation satellites are in a polar orbit. The polar orbit is a low Earth orbit of about 400–900 km altitude in which the Earth observation satellites travel around the planet from the north to the south poles. Typically, but not always, the orbit is Sun synchronous; that is, the plane of the

Table 1.2 Example Revisit Times for Earth Observation Satellites

Satellite	Revisit time
GOES	30 minutes
NOAA	12 hours
Landsat	16 days
ERS	35 days

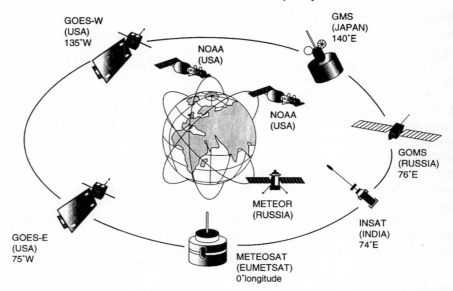

Figure 1.3 Geostationary and Polar Orbits used in Earth Observation. The Example shows the Global Meteorological Satellite System Consisting of Geostationary and Polar Orbiting Spacecraft. Source: EUMETSAT (1991), reproduced by permission of EUMETSAT

orbit of the satellite is in the plane of the Sun and in effect the Earth spins round underneath the satellite as it orbits the planet. This has the advantage of enabling the satellite to revisit a particular region of the Earth at the same time of day.

Instruments

Earth observation instruments can be grouped into imaging instruments and profiling instruments, and their mode of operation into passive and active. The most common type of imaging instrument is the passive radiometer, such as those used on the SPOT, Landsat and NOAA missions, which collects radiation in selected wavebands for each instantaneous field of view the instrument senses. There are also active imaging instruments, such as a synthetic aperture radar, which provides its own source of energy and measures the quantity of energy returned to the instrument by the area being sensed. The imaging instruments are used for observing the land, ocean and ice surfaces of the Earth and for measuring clouds.

The profiling instruments are concerned with measuring the vertical profile of the atmosphere for characteristics such as temperature, moisture, ozone concentration and trace gas concentrations. The passive profiling instruments operate as sounders while the active instruments are profiling radars and lidars which emit their own source of radiation and measure the return radiation.

4.3 Satellite Missions

Introduction

The purpose of this section is to describe the main satellite Earth observation missions. The descriptions are presented in alphabetical order by country or region. The main missions or mission series are given and they cover both current and future activities (CEOS 1995). In the case of future missions, not all are firmly approved, but the descriptions show the plans of the Earth observation data suppliers and the trends in the Earth observation investments.

Table 1.3 lists example World Wide Web (WWW) sites where more information can be gained on the Earth observation missions.

Brazil

The *CBERS* mission is the responsibility of the Brazilian space agency (INPE) working jointly with the Chinese. It is equipped with cameras and scanners for environmental monitoring of the land surface.

Table 1.3 Sources of Earth Observation Data and Information Available at WWW Sites. Source: CEOS (1996)

Country or region	WWW address	Description
Canada	www.ccrs.nrcan.gc.ca/gcnet	Canada Centre for Remote Sensing and Canadian global change research network
Europe	www.ceo.org	European Commission Centre for Earth Observation
	gds.esrin.esa.it:80	ESA Guide and Directory Service
France	www.spotimage.fr	SPOT programme
International	gds.esrin.esa.it/CCEOSinfo	CEOS Information System
Japan	www.eoc.nasda.go.jp	Japanese space agency
United States	harp.gsfc.nasa.gov	NASA Earth Observation System
	www.gcrio.org	US Global Change Research Information Office
	www.saa.noaa.gov	NOAA Satellite Active Archives

Canada

The Canadian Space Agency (CSA) has launched the polar orbiting *Radarsat* satellite. As its name suggests, the satellite carries a radar which is intended for ice, ocean and land applications. Radarsat has a strong operational and commercial orientation.

China

As well as joint work with Brazil on CBERS, China has a programme of *FY* satellites for meteorology from geostationary orbit using a three-channel radiometer, and for meteorology and environmental monitoring from a polar orbit using a ten-channel radiometer.

Europe

The European Space Agency has launched two polar orbiting Earth observation satellites in its *ERS* series. The missions focus on global and regional environmental data collection using active microwave instruments plus the Along Track Scanning Radiometer (ATSR) for sea surface temperature mapping and the Global Ozone Monitoring Experiment (GOME) on ERS-2 for atmospheric chemistry measurement.

The Earth observation satellite after ERS will be the *Envisat-1* mission, which is a multi-instrument mission to follow on from the ERS measurements and to add more atmospheric chemistry and ocean instruments. Envisat-1 will also be in a polar orbit.

ESA is evaluating missions named *Earth Explorer* and *Earth Observer*. These missions will concentrate on science and monitoring respectively: Earth Explorer missions will act as proof of concept missions or platforms for collecting specific Earth observation data, while Earth Observer missions will focus on operational data collection.

EUMETSAT is the European organisation responsible for the Meteosat series of geostationary weather satellites. Meteosat carries radiometers that sense the Earth's disc beneath the satellite in wavelengths from the visible to the thermal infrared. The current Meteosat satellites will be followed by the Meteosat Second Generation series with greater spatial, spectral and temporal resolutions.

ESA and EUMETSAT are collaborating on plans for a *Metop* satellite. Metop will be polar orbiting and will collect data for meteorology and for climate studies, including atmospheric profiles, radiation budget measurements and cloud and surface temperatures. On the EUMETSAT side, Metop is termed the *EUMETSAT Polar System* (EPS) as the first of a potential European operational series of polar orbiting satellites to collect data on the atmosphere.

Introduction

France

The *SPOT* satellite series provides high resolution data primarily of the land surface in visible and near infrared wavebands. The three SPOT satellites launched so far are polar orbiting and are equipped with CCD sensors capable of capturing data with pixel sizes of 10 and 20 m. SPOT 4 will also be equipped with a low resolution sensor to monitor global vegetation (Arnaud 1995) and SPOT 5 will have a very high spatial resolution sensor.

France is a participant with NASA in the *TOPEX/POSEIDON* mission to measure ocean topography using a radar altimeter, and hence the speed and direction of ocean currents.

India

The Indian Space Research Organisation (ISRO) has a series of *IRS* satellites to provide medium to high spatial resolution data of the Earth's surface. The strategy for IRS has three strands: very high spatial resolution instruments with pixel sizes as small as 1 m, instruments for global change research and radar instruments. Initially the IRS polar orbiting satellites were used to collect data mainly of interest to Indian users, but the data are now more widely available to the international community through collaboration with companies in Germany and the USA.

The Indian contribution to the global geostationary weather satellite system is the *INSAT* series of satellites. As with all these satellites, the data collected are primarily full disc images in the visible, near infrared and thermal infrared wavebands.

Japan

Japan makes its contribution to the global geostationary weather satellite system through its *GMS* satellites. The *MOS* polar orbiting satellite operates in the visible, near infrared, thermal infrared and microwave wavebands to take measurements mainly of the ocean.

Japan also provides polar orbiting data through its *JERS-1* satellite, which has on board a radar instrument and an optical instrument, and its *Advanced Earth Observation Satellite* (*ADEOS*), which is a multi-instrument satellite system for physical oceanography, atmospheric dynamics, water and energy cycles, atmospheric chemistry and high spatial resolution land monitoring. An advanced land satellite, *ALOS*, is planned for high spatial resolution land monitoring.

Japan is participating with NASA on the *Tropical Rainfall Monitoring Mission* (*TRMM*) satellite. TRMM is designed to measure tropical rainfall, particularly over the Pacific Ocean, and so contribute to a better understanding of atmospheric dynamics and the hydrological cycle.

Russia

Russia's Earth observation activities can be divided into three main strands. The geostationary *GOMS* and the polar orbiting *METEOR* series provide data for operational meteorology and for climate science. The *Okean* series measures oceanographic and hydrometeorological variables, and the *Resource* series have high resolution sensors for land applications including crop, soil, forest and pollution monitoring. In addition to the Resource series, the *ALMAZ* missions provide radar data of land surfaces and the oceans.

Russian sources also make available very high resolution data from space, which are comparable with air photography. The pixel size of these data is as high as 2 m and they are marketed by an organisation named WorldMap. The data originate from a variety of sources, including the digitisation of pictures taken by spaceborne cameras.

Sweden

In addition to being a minority investor in the SPOT satellite system, Sweden is launching the *Odin* Earth observation satellite for atmospheric studies. In addition to visible and thermal infrared wavebands, Odin will carry an ultraviolet sensor in a polar orbit.

Ukraine

Ukraine has a long history of involvement in space (Krawec 1995). Ukraine's *SICH* series of satellites was originally designed to collect data on physical oceanography and hydrometeorology and then expand to include measurements of the land surface including crop and soil monitoring, forest and tundra fires and pollution monitoring. Ukraine collaborates with Russia on the Okean satellites.

USA

The USA is the pre-eminent Earth observation nation. In 1960 the USA launched the first weather satellite, and in 1972 it launched the first land resources satellite. The two main organisations for Earth observation in the USA are the National Aeronautics and Space Administration (NASA) and the National Oceanic and Atmospheric Administration (NOAA). NASA is responsible for research and development in space activities, while NOAA is responsible for the operational programmes that follow the research stage.

NOAA is responsible for both the geostationary *GOES* meteorological satellites and for the polar orbiting weather satellites, also termed the *NOAA* satellites. There are two GOES satellites to cover the Americas and parts of the Pacific and Atlantic Oceans, both collecting data on the full Earth disc in wavebands covering the visible to the thermal infrared. NOAA satellites have a similar range of wavebands.

Both NASA and NOAA have had a part to play in the *Landsat* series of Earth resources satellites. Landsat, a polar orbiting satellite with wavebands in the visible, near infrared, middle infrared and thermal infrared, has been a pathfinder for high spatial resolution Earth observation and will continue as a mission line within the US Mission to Planet Earth (MTPE) programme.

The US has an interest in oceanographic satellites through the *Seastar* system, which carries a wide field of view instrument for oceanographic measurements, and a continuing involvement in measuring the chemistry and physics of the upper and lower atmosphere through the *UARS* and the *TOMS* systems.

NASA has launched several *Shuttle* missions with Earth observation payloads, including the series of imaging radars and optical systems that have been used to collect data on specific targets.

The next major Earth observation activity of the USA is the *Earth Observing System (EOS)*. Originally conceived as the US Polar Platform, the EOS programme is now divided into the following suite of polar orbiting missions. Many of the activities are joint with other nations (CES 1995).

- *EOS-AERO*: atmospheric chemistry and aerosol properties
- *EOS-ALT*: radar and laser altimeters for oceanography, land and ice surface mapping
- *EOS-AM*: atmospheric dynamics, chemistry and physics, atmosphere–land energy exchanges, vertical profiles
- *EOS-CHEM*: atmospheric chemistry
- *EOS-COLOR*: ocean colour, ocean biology, role of the ocean in global carbon and biogeochemical cycles
- *EOS-PM*: atmospheric properties including cloud physics, atmospheric–ocean energy fluxes, sea ice extent and vertical atmospheric profiles

NASA and NOAA are public sector organisations. In the private sector, the 1994 presidential directive on Earth observation has freed up the ability of US industry to exploit very high resolution optical Earth observation systems with pixel sizes as small as 0.82 m. Companies such as Lockheed Martin and Ball Aerospace have developed missions to collect very high resolution land surface data from a polar orbit. Several jointly owned companies have been formed to launch the *Space Imaging*, *Orbimage* and *EarthWatch* very high resolution missions.

5 DATA POLICY

Earth observation is an international sector where collaboration is important to reduce the costs of expensive space missions and to maximise the value of the investments in Earth observation.

Earth observation is still dominated by governments, although there is an interesting growth of private sector initiatives through the new very high resolution systems. The government dominance means that inevitably government investment is influenced by broader government policies and so decisions on the conditions of access to data reflect political ideologies as well as technical capabilities.

Data policy is important and even vital to the development and maturity of the Earth observation sector. Increasingly the issues of access to Earth observation data are seen internationally to be as important as the technology of Earth observation.

This book discusses the pressures that are being exercised to influence, form and guide the development of the Earth observation sector. Developing the correct data policies will encourage a greater maturity of the Earth observation sector and its sustainability. Developing data policies that are sorely in conflict or that damage the sector internationally may have disastrous consequences.

2
Influencing Factors

1 INTRODUCTION AND PURPOSE

The purpose of this chapter is to examine the factors that influence the development of Earth observation data policy. Initially the chapter discusses the objectives for Earth observation data policy and the growth of data supplies and major user programmes; this is followed by an assessment of the environmental and economic value of the data and the link to market development.

2 OBJECTIVES FOR EARTH OBSERVATION DATA POLICY

The organisations involved in Earth observation are wide ranging in number and in nature. Governments, scientists, operational users and the commercial sector are all part of the Earth observation sector, and all have a voice on the subject of Earth observation data policy. This section describes the objectives which these different parts of Earth observation want to achieve from Earth observation data policy, and how they adapt to conflicting situations.

2.1 Objectives of Governments and Nations

First among the discussion of objectives are those objectives which governments and nations want to achieve in Earth observation data policy. This is different from what nations want to achieve by investing in Earth observation, but the forces which they bring to bear on Earth observation data policy are directly related to their own expectations and ambitions in Earth observation. All governments investing in Earth observation see a benefit in observing the planet from space. This benefit may be short term and result in an immediate market for the data. Other benefits are seen as long term, leading to self-sustaining and mature markets for Earth observation or to the provision of data sets on our environment which can be used for the public good.

There is something of a polarity of views on Earth observation data policy. The US and Japan see investment in Earth observation as part of a social investment to deliver a public service of environmental information. The Europeans acknowledge this public benefit element, but European governments also wish to see a financial return on their investment. For them the issue is not just improved environmental information, but also how to develop a mature market and create the conditions for a government exit from state-funded research and development in Earth observation.

The US Global Change Research Programme (USGCRP), which invests significantly in Earth observation, is a good example of the US public-good information service perspective. The planned budget of the USGCRP in financial year 1997 is US$1.73 billion and the 'underlying premise of the USGCRP is that appropriate use of the Earth for human habitation and survival is inextricably linked to an improved understanding of the systems that are undergoing change in response to natural and human-induced processes' (NSTC 1996). The justification revolves around the human habitation of the Earth and the survival of the human species, and furthermore, 'This broad approach recognises the profound economic, social, and ecological implications of global changes and the need for US leadership in this area' (NSTC 1996).

Similarly Japan, through its ADEOS and TRMM programmes in particular, is contributing data to the improvement in our understanding of global change by an investment in Earth observation as a publicly funded service.

European governments see the benefit of observing the Earth from space for the collection of global change data (ESA 1994a,b), but in addition there is a concern for how to achieve a sustainable Earth observation sector that is not dependent on government sponsorship funding for its future existence. This term needs some explanation. One of the major contributors to ESA's Earth observation programme is the UK. The UK contribution to Earth observation in ESA is provided mainly from the Department of Trade and Industry (DTI). The DTI itself is not a user of the data, it has no programmes which need Earth observation data, but acts as a sponsoring department thereby providing UK government funds to Earth observation in a sponsorship mode rather than in a user or customer mode.

The role of the European Space Agency is crucial to the development of Earth observation in Europe (Harris 1995). When ESA was established in May 1975, the European space delegates resolved that, 'The programmes of the European Space Agency must facilitate the development of operational systems that would be acceptable to and operated by users.'

This orientation to operations was reinforced at the European space ministers meeting in Spain in 1992 where the long-term plan of ESA included the following policy: 'To preserve the Agency's orientation to research and development and to concentrate the resources available to it in these fields, the European space programme shall promote the transfer of mature space systems to operational entities.'

ESA has achieved this transition to operational status with the Meteosat programme and with the Intelsat and Inmarsat communications satellite programmes. Although little progress has been made in developing Earth observation as an operational sector, ESA has a clear mandate from its member states to follow this route. The European Commission and EUMETSAT, the recipient and now operator of the Meteosat operational programme, also see the value for Earth observation in Europe of a transition to operational status.

This perspective is not uniform throughout Europe, and many smaller states see ESA as their Earth observation research and development arm, but increasingly as European nations are under greater budget pressures they are seeking to identify either financial returns from space or ways of ending their research sponsorship funding of Earth observation.

2.2 Objectives of the Scientific Sector

It has long been recognised (Barrett and Curtis 1982) that Earth observation provides many benefits to the environmental sciences. Barrett (1974) and Harris (1987), for example, identify the following major benefits of Earth observation:

- improved data coverage, particularly in remote areas and for large regions;
- homogeneity of data collection by the use of a single instrument to capture data over large areas;
- high frequency of data collection;
- using imaging sensors, Earth observation data are spatially continuous;
- low cost compared to ground surveys.

The core of the benefits for the scientific sector is the ability of Earth observation to provide independent and consistent measurements of environmental phenomena. Measurements from space have the unique advantage of providing continuous, global coverage with high temporal and spatial resolution (ESF 1992).

While the benefits can easily be listed for SPOT and Landsat data, they are valid also for non-imaging systems. The recent developments in climate research owe much to the independent and synoptic perspective of satellites such as the Upper Atmosphere Research Satellite (UARS).

What the scientific sector wants from Earth observation data policy is the opportunity to achieve these benefits. The outcome of research activity is scientific progress, which is achieved through publication, evaluation and dissemination of results, and the scientific sector wants Earth observation data policy both to enable and to encourage greater scientific endeavour.

2.3 Objectives of the Operational Sector

The motivation of operational agencies is quite different from other organisations. Operational agencies, such as NOAA and EUMETSAT, are driven by an operational mandate from a higher level body and, in the case of Earth observation, this operational mandate carries a requirement for information. The key issue is to define an operational information product and then to secure a long-term, continuous supply of that information product. While there may be changes in the instrument behaviour over time, or indeed different instruments may supply the raw data, the central issue is to retain the final information product specification unchanged. Therefore, while Meteosat satellites may change in orbit as new missions are launched, or the NOAA satellite orbit drifts and crosses the equator at a later time over a period of years, the operational agencies need to maintain integrity of their information products by inter-calibration.

For the operational sector the requirement of Earth observation data policy is to provide the conditions for stability and continuity. Short, experimental missions are of some interest to such organisations but not as the provider of regular and stable information product supplies. Changes in policy have as great an effect on operational organisations as do changes in instrument design.

2.4 Objectives of the Commercial Sector

In the commercial sector there are three types of organisation. First, there are the space segment manufacturers and their associated parts companies. Second, there are the companies which act as data brokers, provide value added services and build data processing and analysis systems. These companies are probably the largest in number at present in the commercial Earth observation sector. Third, there are the end-users. The end-users are like the operational sector in that they are more interested in the final information product and how it assists in better decision making than in the details of the original data and the data processing chain.

The objectives of the first two groups in the commercial sector are to provide space products and services efficiently and at a profit. These organisations also want the expansion of the Earth observation sector so that the market for products and services is larger. They want Earth observation data policies that encourage market expansion, and do not want to have policies that allow preferential access to organisations not equipped to provide long-term information supplies.

The objective of the end-user group is to use Earth observation information as part of their own value chain to achieve greater efficiencies in their business processes, which typically means driving down the cost of those processes. Information products derived from Earth observation can help business to be better, cheaper or faster.

2.5 Coping with Conflict

There is the potential for conflict on data policy between organisations with objectives other than Earth observation, and between those in different sectors within Earth observation. The conflict typically relates to belief systems regarding the reason for state investment in space. Here there is a distinction between government investment as a sponsor and government spending as a customer. In the US and Japan the case has been made and the argument won for government investment in Earth observation as a means to the end of better information for environmental and global change research (NSTC 1996). This means that these governments go on funding Earth observation as sponsors for the public benefit of improved understanding of the planet Earth.

In Europe that case is not so strong and there is an equally loud voice claiming the need for programmes to have clear ways for customers to fund Earth observation programmes in the long term. This does not mean that governments cannot provide funds for Earth observation, but that they do so as customers for information and not as sponsors of the sector.

It could be argued that it is up to individual organisations to set their own data policies irrespective of the actions of others, but international interdependence in Earth observation is a major characteristic of the sector. A report of the US Committee on Earth Studies (CES), part of the National Research Council, notes the international interdependence of US Earth observation on several occasions:

> In sum, the US national needs for data to support research in the earth sciences or for operational earth applications are absolutely reliant on international collaboration and the continued availability of data from foreign countries. An early precept of the US space program was that the United States should always retain the capability to conduct a program on its own, even though it might enter into a cooperative international venture. While that precept held sway for much of the history of the space program, it has long since evaporated in the conduct of earth observations – and properly so. (CES 1995: 17)

> Except for Shuttle-based Synthetic Aperture Radar (SAR), the United States has ceded its planned work on SAR to the Europeans, Japanese, and Canadians. (CES 1995: 37)

> For the more robust, free-flyer missions, the United States is now reliant upon data exchange agreements to satisfy US needs. (CES 1995: 37)

> At $8 billion, EOS must depend increasingly on our European and Japanese partners. Failure to accomplish planned international cooperation ... will leave gaping holes in

the international Earth Observing System. We note that in the $8 billion program the US is relinquishing to international partners the development of new advanced technologies in laser and active microwave remote sensing. (EOS Payload Advisory Panel 1992; quoted in CES 1995: 49)

The activities of the Committee on Earth Observation Satellites (CEOS) are in part a response to this need for interdependence and for international agreement to share burdens of investment in Earth observation yet provide access to complementary Earth observation data.

There are also conflicts between the scientific sector and others on Earth observation data policy. The scientific sector often sees Earth observation data as a public good because it is information about our environment, the capture of which has been mainly funded by governments. In turn there is a perception that Earth observation data should be free or available at low cost (ESF 1992). If data and information products produced by the Earth observation sector are provided at low cost then the funding for future missions may be harder to secure from the sector itself.

3 GROWTH OF DATA SUPPLIES

There are two main growth factors that are having an important influence on the Earth observation sector. One key factor is the growth of data supplies through time. We are currently experiencing a dramatic increase in Earth observation data as part of a sharply increasing curve in both the quantity and the diversity of data. A second factor is the growth and maturity of global and regional research and monitoring programmes which use or will use Earth observation data as key parts of their programmes.

The first weather satellite was launched in 1960, the first Landsat satellite was launched in 1972 and the first SPOT satellite in 1986. However, it has been during the 1990s that we have seen a dramatic growth in Earth observation data supplies. In 1991, ESA successfully launched its ERS-1 satellite, followed by ERS-2 in 1995, and in 1992 Japan launched its J-ERS-1 satellite. During the 1990s there has been a substantial increase in the supply of Earth observation data (Logica 1991; Harris 1992) and this trend will continue through to the next century with the launch of the US Earth Observing System satellites, ESA's Envisat-1, Japanese missions such as TRMM and ALOS, and the Earth observation missions of emerging space nations such as India and Argentina.

In parallel with an increase in data and an increase in the international base of Earth observation, there is also an increase in the variety of instrument types. The early satellite programmes consisted of spacecraft with single or small numbers of instruments whose data were intended for a variety of applications. Developments in the 1990s comprise dedicated missions with small numbers of instruments for specific applications (e.g. Radarsat for ice monitoring and Space Imaging for mapping), and large, multi-instrument platforms such as Envisat-1 designed for a wide variety of applications. In the dedicated mission class there is a growing interest in small satellites, so-called smallsats or lightsats, which can be constructed cheaply and used to fly smaller instruments; an example is the SAC-C mission of the Argentinian space agency.

The *1995 CEOS Yearbook* illustrates the great growth in the number of instruments for Earth observation that have been or will be launched in the period to the year 2009 (CEOS 1995). Figure 2.1 shows the durations of the approved or planned Earth observation missions taken from the *CEOS Yearbook*. These missions relate to the members of CEOS and so do not include the private sector initiatives, such as the US very high resolution systems, which will add further to the volume of data supplies.

It is foreseen by CEOS (1995) that by the year 2009 there will be 126 missions flying 217 instruments for Earth observation from the many and growing number of nations active in space. Collectively these instruments will provide approximately 20 terabits (20×10^{12} bits) of data per day during the early part of the next century (Harris 1992). The volume of this vast amount of data is hard to visualise: as a comparison it is the equivalent of the data content needed to create 600 000 bibles per day.

The growth in the amount of data is important for data policy. The pricing policies applied to these data will be a strong influence on the extent of the use of the data. If the pricing policy and conditions of access are right for the user community then these policies will encourage maximum use of the data. If they are not right then the development of the Earth observation sector will be held back.

The growth in the volume and diversity of data will also have implications for data reception, dissemination and archiving. As the data policy of the International Geosphere Biosphere Programme notes, it will be possible to store all the new Earth observation data, but a key question concerns the management of the stored data (Townshend 1996).

4 GROWTH IN FORMAL USER PROGRAMMES

4.1 Impetus

One of the responses to a growing worldwide concern for the status of the environment has been a growth in the number of formal scientific research and monitoring programmes. These programmes are a response to concern over the sustainable development of the planet Earth, and have been

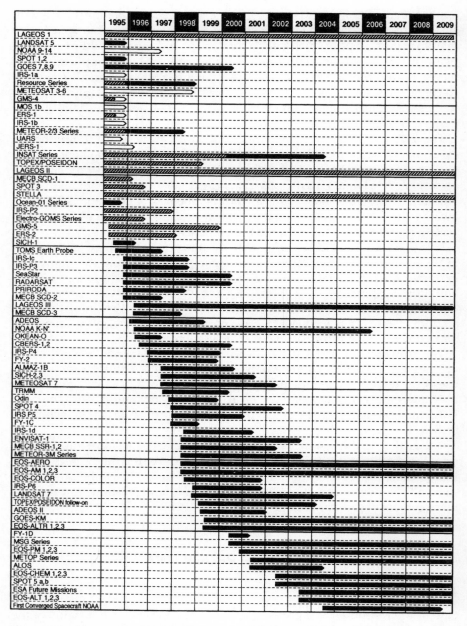

Figure 2.1 Summary of Earth Observation Missions, 1995 to 2009. Source: CEOS (1995). Reproduced by permission of the European Space Agency

Influencing factors 29

reinforced by a range of international environmental agreements including the Framework Convention on Climate Change, the International Convention to Combat Desertification, and the Montreal Protocol. Each of these agreements explicitly calls for systematic observation of the Earth to increase our understanding of its physical processes and our ability to monitor them.

The purpose of this section is to describe the major scientific and operational programmes that require large-scale, systematic environmental observation and which therefore are and will be heavy users of Earth observation. These programmes influence the development of Earth observation data policy, and the use they make of Earth observation data is itself influenced by the data policies. The discussion in this section is split into two themes: global science and global monitoring.

4.2 Global Science

International Geosphere Biosphere Programme

The International Geosphere Biosphere Programme (IGBP) was established by the International Council of Scientific Unions (ICSU) in 1986 (IGBP 1992). The IGBP addresses the dynamic and complex nature of the Earth system, its past changes and its future development. The aim of the IGBP is as follows: 'To describe and understand the interactive physical, chemical and biological processes that regulate the total Earth system, the unique environment that it provides for life, the changes that are occurring in this system, and the manner in which they are influenced by human activities' (IGBP 1992).

Since the research area is so vast, the IGBP does not provide an umbrella for all relevant research but comprises a number of core projects which target high-priority scientific questions. The IGBP has chosen six key questions which flow from its main objectives. These questions are shown in Table 2.1 together with the core project(s) that relates to each question.

The IGBP has recognised the great demand for Earth observation data and has established as one of its main activities the IGBP Data and Information System (IGBP-DIS) to cover the IGBP data needs. The IGBP-DIS is not a central data archive but rather a means of encouraging and facilitating a wider availability of data, including Earth observation data, which are useful for global change researchers in the IGBP community.

World Climate Research Programme

While the IGBP is concerned with the interactive physical, chemical and biological processes of the Earth system, a separate major project is concerned with the physical climate system. Since 1979, ICSU and the World Meteorological Organisation have jointly undertaken the World Climate

Table 2.1 The Six Key Questions of the IGBP and the Corresponding IGBP Core Projects. Source: IGBP (1992)

IGBP question	Core project(s)
How is the chemistry of the global atmosphere regulated, and what is the role of biological processes in producing and consuming trace gases?	International Global Atmospheric Chemistry (IGAC)
How will global changes affect terrestrial ecosystems?	Global Change and Terrestrial Ecosystems (GCTE), Land Use/Cover Change (LUCC)
How does vegetation interact with physical processes of the hydrological cycle?	Biospheric Aspects of the Hydrological Cycle (BAHC)
How will changes in land use, sea level and climate alter coastal ecosystems, and what are the wider consequences?	Land–Ocean Interactions in the Coastal Zone (LOICZ)
How do ocean biogeochemical processes influence and respond to climate change?	Joint Global Ocean Flux Study (JGOFS)
What significant climatic and environmental changes occurred in the past, and what were their causes?	Past Global Changes (PAGES)

Research Programme (WCRP), and they have now been joined by the Intergovernmental Oceanographic Commission (IOC) as a third lead body (Perry 1993).

The WCRP seeks to determine two elements of our climate. First, to what extent can climate be predicted? Second, how are human activities affected by climate? The WCRP focuses on the physical and dynamic aspects of these problems and studies how climate behaves given the composition of the atmosphere and the character of the land surface.

As with the IGBP, the WCRP is organised into a series of research projects. These projects are concerned with: the characteristics of the oceans and the related water circulation; processes within the atmosphere including specifically the stratosphere; climate variability and predictability; and the Arctic climate system.

Human Dimensions Programme

The third science programme is concerned with the human role in global change and the implications of global change for human society. The Human Dimensions of Global Environmental Change Programme (HDP) concentrates on the human elements of global change rather than the physical, chemical and biological elements.

The HDP has produced a research agenda which consists of a set of four questions. The four questions (shown in Table 2.2) recognise that human activities on the planet have a link both with physical processes such as those studied within IGBP and WCRP, and institutional structures, social organisation and cultural traditions. Accordingly, the HDP has identified seven research areas (also shown in Table 2.2) that will help to answer the four broad questions.

Data and Data Policy for Science Programmes

Each of the three major science programmes has a similar organisation, namely a core set of internationally agreed objectives plus a group of selected projects focused on particular themes. In all of these programmes, Earth observation is an important source of data. For example, both the IGBP and the HDP have a strategic concern for land use and land cover changes. A major source of data on land changes is Earth observation, and the issues of access to these data become central to the production of the scientific results of the programmes.

Because the IGBP, the WCRP and the HDP are global in reach, the Earth observation data sets can in many cases be very large. Therefore, both for reasons of their scientific agendas and the size of the data sets involved, the issues of Earth observation data policy are important for the development of the IGBP, the WCRP and the HDP.

Table 2.2 The Main Questions and Research Priorities of HDP. Source: HDP (1994)

HDP questions	Priority research areas
• What are the human driving forces of global change?	• Land use/cover change • Industrial transformation and energy production and consumption
• How are these social processes linked to the physical processes of global change?	• Population growth and social change • Demographic and social dimensions of resource use
• What are the impacts of physical global change processes on human systems? • What is the potential for adaptation and what mitigation strategies could different social systems formulate and adopt?	• Influence of institutional structures • Environmental security and sustainable development • Attitudes, perceptions, behaviour and knowledge

4.3 Global Monitoring

Global Climate Observing System

During the 1980s and 1990s there has been a growing concern over global climate change. The dimensions of the concern and the progress in assessments of climate are best summarised in the work of the Intergovernmental Panel on Climate Change (Houghton *et al* 1995). The assessment by the IPCC identified both the uncertainties in climate prediction and the parts of the climate system for which a better understanding is required.

To address these limitations a call for coordinated action to measure the climate system was developed at the Second World Climate Conference in 1990 and reaffirmed at the United Nations Conference on Environment and Development held in Rio de Janeiro in 1992. The ministerial declaration at the Second World Climate Conference noted that,

> to reduce uncertainties, to increase our ability to predict climate and climate change on a global and regional basis, including early identification of as yet unknown climate-related issues, and to design sound response strategies, there is a need to strengthen national, regional, and international research activities in climate, climate change and sea-level rise (SWCC 1990).

A response to this concern has been the establishment of the Global Climate Observing System (GCOS). The four sponsoring organisations of GCOS are as follows:

- Intergovernmental Oceanographic Commission of UNESCO (IOC)
- International Council of Scientific Unions (ICSU)
- United Nations Environment Programme (UNEP)
- World Meteorological Organisation (WMO)

GCOS has been established to develop a dedicated observation system designed to meet the scientific requirements for monitoring the climate, detecting climate change and predicting climate variations and climate change (GCOS 1993).

The early priorities for GCOS are to coordinate and facilitate the observational tasks needed to address the main climate issues such as seasonal and interannual climate prediction, early detection of climate trends and climate change caused by human activity, and a reduction in the major uncertainties in climate prediction. In order to do this, GCOS has developed a strategy that includes *in situ* data, satellite data and derived information. Because spaceborne Earth observation data are the largest in volume of these data, the data policy issues which apply to the data take on considerable importance.

Global Ocean Observing System

The uncertainties in our knowledge of the oceans are great, and yet the major conditioning factor for future climate is the state of the oceans. The transport of heat, water, gases and nutrients by the oceans largely determines future weather and climate in the medium and long term.

Despite the importance of the oceans, the information available is relatively limited. As a response to this limitation a Global Ocean Observing System (GOOS) has been proposed by a similar group of partners to GCOS, namely the Intergovernmental Oceanographic Commission, the United Nations Environment Programme and the World Meteorological Organisation. The Global Ocean Observing System has been described as 'a scientifically designed permanent, international system for gathering, processing, and analysing oceanographic observations on a consistent basis, and distributing data products' (GOOS 1993).

GOOS will build on existing and planned scientific and observational programmes such as TOGA (Tropical Oceans and the Global Atmosphere), WOCE (World Ocean Circulation Experiment) and GEWEX (Global Energy and Water Cycle Experiment). GOOS will gather Earth observation data and *in situ* data from the ocean surface and the ocean depths in order to respond to developing demands for ocean data. The Earth observation data recognised to be valuable to GOOS are particularly from satellite altimeters, wind scatterometers, ocean colour sensors and sea surface temperature instruments.

Global Terrestrial Observing System

The final global observing system of the trio of monitoring programmes is the Global Terrestrial Observing System (GTOS). GTOS was established in 1996 in response to international calls for a deeper understanding of global change, particularly as it involves the land surface rather than the atmosphere or the oceans. The central mission of GTOS is 'to provide data for detecting, locating, quantifying and giving early warning of changes in the capacity of terrestrial ecosystems to sustain development and improvements in human welfare' (GTOS 1996).

GTOS has been established by the Food and Agriculture Organisation of the United Nations (FAO), ICSU, UNESCO, UNEP and the WMO, and is oriented to contributing to the answers to five key questions:

- What are the impacts of land use change and degradation on sustainable development?
- Where, when and by how much will demand for freshwater exceed supply?
- Where and when will toxic pollutants cause major threats to human and environmental health and the capacity of ecosystems to detoxify them?
- Where and what type of biological resources are being lost, and will these losses irreversibly damage ecosystems or human progress?
- What are the impacts of climate change on terrestrial ecosystems?

GTOS is envisaged as a partnership of partnerships, formed largely by linking existing monitoring sites and observational networks, plus Earth observation systems to provide for hierarchical measurement of the land surface. The outline sampling hierarchy is shown in Table 2.3.

One of the roles of Earth observation in this GTOS structure is to provide the scaling up from a detailed understanding of local-scale processes to the regional level. SPOT and Landsat data can be used to identify surface features associated with ground measuring sites in detail, and then scaled up using NOAA AVHRR or ERS ATSR data to cover larger regions. These data can then be used in models to contribute to answering the five questions listed earlier.

As with GCOS and GOOS, Earth observation plays a significant role in GTOS and therefore the development of data policies will either hinder or benefit the improved understanding of the processes in the terrestrial environment. This is particularly important given the long-term perspective of GTOS and the need for continuity of data supplies.

GTOS has developed guiding principles for its data management system and, because of the significant overlaps in information requirements, has agreed a common data management plan among GCOS, GOOS and GTOS. As well as the databases of environmental information, the data management plan foresees high-quality data referencing and metadata systems with tools for searching, sorting and exchanging data.

Integrated Global Observing Strategy

GCOS, GOOS and GTOS clearly have overlaps in data requirements and methods of analysis, not least because they are all concerned at some level with climate change. The concept of the International Global Observing Strategy (IGOS) has arisen from an international realisation that greater coherence of observing systems would be more effective. A complete IGOS has not yet been established, but Figure 2.2 gives an overview of the main components proposed for an IGOS.

Table 2.3 The Sampling Hierarchy Planned for GTOS. Source: GTOS (1996)

Tier	Observation type	Number of samples
1	Large-scale experiments and gradient studies	10
2	Long-term research centres	200
3	Field stations	1000
4	Unstaffed sampling sites	10 000
5	Satellite Earth observation	> 100 million

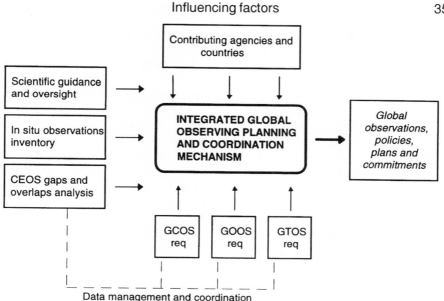

Figure 2.2 A Summary of an Integrated Global Observing Strategy. Source: Anon (1996)

The global observing systems have a strong interest in the long-term development of stable data policies. Their rationale is long-term monitoring and so their requirement is for long-term data supplies that have a consistent technical and data policy basis.

5 ECONOMIC AND ENVIRONMENTAL ISSUES

5.1 Information Markets and Public Goods

Electronic Information

Economic activity relating to the supply, processing and transmission of many types of information held in electronic form is now central to a growing number of economies. Electronic funds transfer and international financial dealing are clear examples of this, and Earth observation data can be seen as another form of electronic information. All forms of electronic information can be packaged as a product (Spero 1982), and Earth observation is increasingly being seen as an information product. The growing use of the Internet is making electronic information more widely available and easier to use.

One important characteristic of information in electronic form is that it cannot be consumed in the same way as, say, a tin of baked beans or a potato. Electronic information can be used or consumed, but this use does not diminish the ability of another user to use the information: there is no destruction of the information, no reduction in the quantity of information and no changes in the quality of the information. Encryption and licence arrangements can limit the technical and legal access to the information, but not the use of the data once access to the information has been agreed.

The economic characteristics of information products derived from Earth observation data can be assessed from the perspective of their importance as resource inputs and as commodities that can be traded. Earth observation data and the derived information products are equivalent to natural resources. The economic benefit of the information derives from the improved decision making that can be effected through the use of the information (Mansell and Hawkins 1992).

Public Goods

As with all geospatial information, Earth observation data display some of the characteristics of a public good (Ordnance Survey 1996). An example of a public good is a lighthouse. A public good has two main characteristics: non-rivalry and non-excludibility (Pearce 1995). Non-rivalry means that the consumption of the information (e.g. the light from a lighthouse seen by a sailor at sea) does not diminish the capability of another consumer (another sailor at sea). Non-excludibility means that no one user can be excluded from using the information by another user. Using the lighthouse analogy, the group that constructed the lighthouse cannot limit the use of the light from the lighthouse to only those funding its construction. In environmental terms, similar arguments can be rehearsed, for example in relation to air quality.

In areas of global change science and monitoring, there is a strong public-good case. Progress in the science of global change and the widespread dissemination of the results will help all people on the planet through the following:

- provision of information on the state of the planet
- the development of national and international policies based on the results of global change research
- action by producers and consumers based on the policies. Such action will respond to the results of the global change science

The costs of generation, distribution and use of Earth observation information is borne in large part by the public sector as a result of public policy. This mixes with the public good nature of Earth observation data to create tensions over the conditions of access to Earth observation data. The perception that Earth observation data are a free public good (comparable to the light from a lighthouse) is restricting the development of the sector because it has a limited capability to recover the investment costs. The sector is characterised by a mix of public and private sector institutions with differing expectations.

Value Chains

The mix of public and private sector participation in Earth observation can be seen in the value chain of Earth observation shown in Figure 2.3. Earth observation data policies affect all parts of the value chain. The upstream sections are characterised by the public sector (e.g. ESA, NASDA and NASA) because of the high investment costs in the space segment and in the establishment of data processing and the provision of information services. The downstream sections are increasingly under the control of the private sector (e.g. the National Remote Sensing Centre in the UK or Euromap in Germany), although their actions are strongly influenced both by ownership and by the data policies of the upstream section.

The substantial involvement of the public sector as both producer and user of Earth observation data creates an environment in which it cannot be assumed that prices will signal investment priorities that match the priorities of government policy. Market distortions created by the characteristics of information products, the public-goods aspects of Earth observation information and the significant role played by public agencies and their influence on the downstream private sector are major features of Earth observation.

Global, regional and national markets for Earth observation data are quasi-markets. The cost and the price of Earth observation data have been largely determined by administrative judgements (e.g. ESA's pricing policy for ERS-1 and ERS-2 data). This means that there is a balance between administrative and market elements in decision making. An open question which many governments have examined is: to what extent can a private sector element be introduced into Earth observation activities without affecting those services which are intended to serve a public interest objective?

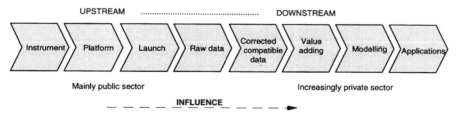

Figure 2.3 Value Chain of Earth Observation

5.2 Environmental Value

Ozone

An illustration of the mix of public and private sector issues is the illustration of the responses to the ozone problem. A significant contribution of Earth observation has been to aid the analysis of the seasonal reduction in stratospheric ozone over the Antarctic region (the so-called ozone hole). The excellence of the scientific work associated with the studies of ozone depletion has attracted negligible economic recognition. However, the value of the discovery, and of the monitoring of stratospheric ozone levels, has an industrial significance measured in billions of US dollars and a health significance measured in trillions of US dollars.

The main cause of the ozone hole has been the widespread use of chlorofluorocarbons (CFCs). Reduced stratospheric ozone, which follows from increased CFC concentrations, produces health problems in humans and damages ecosystems by allowing an increase in ultraviolet radiation to enter the Earth–atmosphere system. Pearce (1995) notes that a 1% reduction in ozone produces a 3–4% increase in non-melanoma skin cancers. In the oceans an increase in ultraviolet radiation affects phytoplankton production by reducing the rate of photosynthesis. The US Environmental Protection Agency (quoted in Pearce 1995) has estimated that for the period 1989–2075 the cost of health and environmental damages could be as high as US$3.5 trillion if there were no reduction in CFC emissions.

To respond to the damage to the atmosphere by CFCs, nations established the Montreal Protocol on Substances that Deplete the Ozone Layer. The Montreal Protocol came into effect on 1 January 1989 and has set agreed targets for CFC reduction.

In the whole process of the identification and monitoring of the problem, and of the political and industrial responses, there has been no substantial financial recognition of the role played by Earth observation data, and yet Earth observation data have been a major component in the recognition and monitoring of the changing ozone hole. There is no process to feed back to the satellite programmes any elements of the financial benefits of policy actions.

Measures

Mostafa Tolba, the Executive Director of UNEP, argued at the UNEP Governing Council meeting in 1990 that 'Placing a monetary value on the component parts of a healthy ecosystem will always present a problem. But difficult as it may seem, we have to refine economic methods to allow every country to begin including environmental auditing and national planning.'

Standard measures of Gross Domestic Product (GDP) and Gross National Product (GNP) do not include measurements of natural capital such as a nation's stock of water, wildlands, soil and non-renewable resources. However, these elements have a value to a nation and damaging these resources has a direct economic cost because it is necessary to carry out alleviating actions.

For Earth observation data policy this presents a problem. Data from Earth observation satellites are particularly good at measuring the Earth's surface, oceans and atmosphere, yet these are the very elements that do not enter into calculations of GDP and GNP. Earth observation data clearly have an environmental value, but this value cannot be measured simply by the market value of the data because the market-based and standard accounting techniques are not set up to capture environmental value (Ahmad, El Serafy and Lutz 1989; Bartlemus, Stahmer and van Tongeren 1991; Lutz and Munasinghe 1991).

The increasing recognition of environmental issues by governments (e.g. the Climate Change Convention and the Biodiversity Convention) is likely to increase the recognition of the need to take greater account of the environment in national accounting practices. As the value of environmental information changes through time, so the implications for Earth observation data policy will change. Policies for environmental protection are likely to become stronger in the future and as Earth observation is one of the main ways to monitor the environment then the policies which govern Earth observation data supplies are likely to change and take on greater importance.

6 COMMERCIALISATION AND MARKET DEVELOPMENT

6.1 Missions

Many Earth observation missions have an objective to develop commercial applications using data from the mission. SPOT and Landsat are the most developed cases of the commercial potential of Earth observation, while the new US very high resolution systems such as EarthWatch and Space Imaging are geared specifically to the commercial sector.

In Europe, ERS-1, ERS-2 and Envisat-1 all have some form of commercial benefit foreseen in their objectives. The ERS data policy, for example, states that 'The purpose of the ERS data policy is to promote the scientific/technological (research) application and commercial objectives of the programme in a balanced way' (ESA 1994c).

In the USA, experience with Landsat has been mixed so far as commercial success is concerned. Landsat had an uncertain start and responsibility for the programme has been moved several times during its lifetime. Table 2.4 lists some of the key activities and their associated dates in the history of the Landsat programme (CES 1995).

So far the experiences in the USA and in Europe of the commercialisation

Table 2.4 A Summary Chronology of the Landsat Programme. Source: CES (1995)

Date	Event
July 1972	Landsat 1 launched. Originally named the Earth Resources Technology Satellite (ERTS) and subsequently renamed
January 1975	Landsat 2 is launched
March 1978	Landsat 3 is launched
May 1978	Presidential Directive 37 is released stating the national policy of encouraging domestic commercial exploitation of space capabilities
November 1979	Presidential Directive 54 assigns responsibility for all civil operational land Earth observation systems to NOAA
July 1981	The Office of Management and Budget states that additional satellites beyond Landsat 5 would depend on private sector investment and operation
July 1982	Landsat 4 is launched
March 1983	NOAA announces the Administration's decision to transfer the United States' civil operational Earth observation satellites, including Landsat and the weather satellites, to the private sector
March 1984	Landsat 5 is launched
July 1984	The Landsat Act is passed by Congress and signed by the President, authorising commercialisation of the US land Earth observation programme
May 1985	Contract awarded to EOSAT for data sales
October 1992	Control of Landsat transferred from the Department of Commerce to the Department of Defense and NASA. The Department of Defense is to procure Landsat 7, and NASA is to manage operations and supervise data sales
October 1993	Landsat 6 is destroyed soon after launch when the upper stage fails to fire correctly
January 1994	The Department of Defense and NASA reach an impasse over funding for Landsat 7, and the Department of Defense withdraws
March 1994	NASA assumes responsibility for Landsat 7
May 1994	A Presidential Directive is issued on Landsat naming NASA the lead development agency and NOAA the operating agency

and market development of Earth observation have not been entirely successful. The evidence in Table 2.4 suggests that the orientation of the Landsat programme has not been clearly defined. Was Landsat a technology demonstration programme (which its original name suggests) or was it part of a definite plan to build a base which the private sector could take over? Much the same questions can be asked in Europe of the ERS missions.

6.2 Stages in Market Development

One way of looking at the development of Earth observation towards a more mature market is presented in a study for ESA by ESYS (1994) and summarised in Figure 2.4. The diagram in Figure 2.4 was developed in the

Influencing factors 41

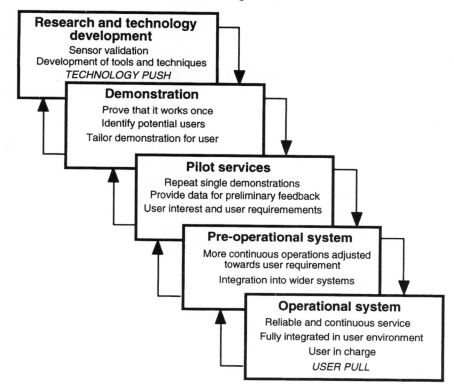

Figure 2.4 Stages in Earth Observation Applications Development. Source: ESYS (1994)

context of improving the number and range of applications of Earth observation, but there are useful lessons more generally for the development of the market for Earth observation data.

Much of Earth observation is at the first two stages in Figure 2.4, i.e. at the stage of developing a technology and demonstrating that it works. It is characterised by technology-push. It is not surprising therefore that the commercial development of programmes such as Landsat and ERS has not been great as they are mainly acting in the first two stages and not acting throughout all stages. The eventual user-pull will not happen without effort, and it is necessary to have clear and coherent objectives across all the stages in Figure 2.4 to achieve strong market development in Earth observation (Smith 1994).

What does this mean for data policy? When organisations develop Earth observation systems that claim to have some form of commercial orientation it is necessary to put in place a chain of activities that assist in market development and not just to rely on data availability as the main contribution. ESA acknowledges the need for balance in scientific and commercial applications of its Earth observation satellites, but appears to act in favour of scientific developments in its data policies.

Much of Earth observation is still at the early stages described in Figure 2.4, yet the importance of the later stages of the diagram are recognised. Data policy can be a tool to promote greater effort in the later stages and so increase the maturity of Earth observation.

3
Existing Policies and Policy Making Processes

1 INTRODUCTION

At global, regional and national levels there has been a widespread development of data policies by organisations which are responsible for or which influence Earth observation systems. The purpose of this chapter is to review the main policy statements that have been produced by organisations working at three levels: global, regional and national. In the discussion of regional Earth observation data policies, the focus is on European-level organisations.

In reviewing the policies themselves the organisations which develop Earth observation data policies are also described. This approach is used to illustrate the policy making processes which create the Earth observation data policies. In general, Earth observation data policy statements are made by two types of organisation. First, there are the organisations which are responsible for Earth observation programmes. They own the data and so can create their own policy statements. Examples of the first group are ESA, EUMETSAT and NASA. Second, there are the organisations which act in a collective manner to identify Earth observation data policy issues that should be considered when data policies are defined in detail. Examples are the International Geosphere Biosphere Programme, the European Science Foundation and the Committee on Earth Observation Satellites.

2 GLOBAL

This first section examines the Earth observation data policies of the three main international organisations that have global or broadly based participation: the United Nations, the Committee on Earth Observation Satellites and the International Earth Observing System.

2.1 United Nations Principles on Remote Sensing

Principles

An important basic legal instrument relating to Earth observation is the set of 15 Principles on Remote Sensing that were approved unanimously by the United Nations General Assembly on 11 December 1986 in its Resolution 41/65 (Jasentuliyana 1988). The Principles have now become customary law and are binding on UN member states.

Remote sensing activities are defined in the UN Principles as the operation of remote sensing space systems, primary data collection and storage stations and activities in processing, interpreting and disseminating the processed data. The scope of the Principles was not intended to be exhaustive and to cover all remote sensing. Principle I limits the applicability of the Principles to sensing of the Earth's surface from space for the purpose of improving natural resource management, land use and the protection of the environment. This has been interpreted narrowly by some as only including the land surface for environmental management purposes. This is too narrow an interpretation. The Earth's surface can include the water and ice surfaces as well as the land surface, and the protection of the environment has wide-ranging consequences for environmental security.

The Principles do exclude the remote sensing of the atmosphere because this is not the Earth's surface, and they exclude military applications because the Principles are concerned only with the civil domain.

Four of the 15 UN Principles on remote sensing are particularly relevant to the questions of data policy addressed in this book (see Jasentuliyana 1988):

Principle II

Remote sensing activities shall be carried out for the benefit and in the interests of all countries, irrespective of their degree of economic, social or scientific and technological development, and taking into particular consideration the needs of developing countries.

Principle V

States carrying out remote sensing activities shall promote international cooperation in these activities. To this end, they shall make available to other States opportunities for participation therein. Such participation shall be based in each case on equitable and mutually acceptable terms.

Principle VI

In order to maximize the availability of benefits from remote sensing activities, States are encouraged through agreements or other arrangements to provide for the establishment and operation of data collecting and storage stations and processing and interpretation facilities, in particular within the framework of regional agreements or arrangements wherever feasible.

Principle XII

As soon as the primary data and the processed data concerning the territory under its jurisdiction are produced, the sensed State shall have access to them on a non-discriminatory basis and on reasonable cost terms. The sensed State shall also have access to the available analysed information concerning the territory under its jurisdiction in the possession of any State participating in remote sensing activities on the same basis and terms, taking particularly into account the needs and interests of developing countries.

These principles are quite general and set a scene of international collaboration in Earth observation, with a particular concern for the needs of developing countries.

Alternatives

The negotiators of the Principles originally discussed two approaches. One approach envisaged that a state responsible for an Earth observation mission would have to gain the permission of all other states for which data collection was required. The second approach, and the one agreed by the negotiators, was to allow a state to observe any other state from space as long as the sensed state could gain access to the Earth observation data collected of its own territory.

Conditions of Access

Based on this agreement, Principle XII is particularly notable in terms of data policy. This principle gives a right to a sensed state to gain access to Earth observation data of its own territory under three main conditions: as soon as the data are produced they will be made available; access to the data by the sensed state will be non-discriminatory; and the access will be on reasonable cost terms. This means that, for example, if Earth observation data for part of Argentina were collected by a European Earth observation satellite then Argentina must have access to the data as soon as they are produced, without any discrimination, and at reasonable cost terms.

The first of these two conditions does not seem to present any substantial problem, but what is meant by reasonable cost terms? It could be interpreted to mean the cost of capturing the data and so mean an assessment of the investment in the European Earth observation mission. Or it could be interpreted as the cost that any other purchaser of data for another part of the world is required to pay. Or it could even be interpreted as a cost related to the ability of the purchaser to pay for the data.

Guidelines

The United Nations Principles were unanimously approved by the United Nations General Assembly in 1986. So far the Principles have not been used in anger by nations or organisations that wish to gain access to Earth observation data. As the issues of access to data grow in importance around the world then there is a greater potential for these Principles to be used in a serious way to challenge Earth observation data suppliers.

Israel is proving to be an interesting case for very high resolution data availability on a geographical basis. For several years some US senators have been concerned about the new US systems collecting data of the state of Israel. The issue has been particularly delicate in the case of Orbimage because the company has an agreement with Eirad of Saudi Arabia to establish a ground station for receiving high resolution Orbview-3 data in Riyadh, Saudi Arabia. Senator Bingaman noted in the US Senate on 26 June 1996 (Congressional Record 1996a) that the US Department of Commerce had agreed that, as a condition of the Orbimage licence, Orbview-3 would not sense the state of Israel. The US Senate has come to the following agreements on the prohibition and release of satellite imagery of Israel (Congressional Record 1996b).

(a) Collection and Dissemination: A department or agency of the United States may issue a license for the collection or dissemination by a non-Federal entity of satellite imagery with respect to Israel only if such imagery is no more detailed or precise than satellite imagery that is available from commercial sources.
(b) Declassification and Release: A department or agency of the United States may declassify or otherwise release satellite imagery with respect to Israel only if such imagery is no more detailed or precise than satellite imagery of Israel that is available from commercial sources.

If it is applied in practice this geographical constraint may be a precedent for other parts of the world.

2.2 Committee on Earth Observation Satellites

CEOS Membership

The Committee on Earth Observation Satellites (CEOS) was created in 1984 in response to a recommendation from the Economic Summit of Industrialized Nations Working Group on Growth, Technology and Employment Panel of Experts on Satellite Remote Sensing (CEOS 1994). The members of CEOS recognised the multidisciplinary nature of satellite Earth observation and the value of coordination across all missions.

CEOS now has 36 members, consisting of 22 full members, 4 observers and 10 affiliates (CEOS 1995). The list of CEOS members is given in the appendix. CEOS works on a best-efforts basis and it has the following three primary objectives (CEOS 1995):

- to optimise the benefits of spaceborne Earth observation through cooperation of its members in mission planning and in the development of compatible data products and services;

- to aid both its members and the international user community by serving as a focal point for international coordination of space-related Earth observation activities;
- to exchange policy and technical information to encourage complementarity and compatibility among spaceborne Earth observation systems.

CEOS Context

The members of CEOS have agreed two sets of Earth observation data policy principles. At the sixth CEOS Plenary meeting held in London in December 1992, members agreed a set of satellite data exchange principles in support of global change research. Subsequent to this, CEOS members in April 1994 in Washington DC agreed to a preliminary resolution on the principles of satellite data provision in support of operational environmental use for the public benefit.

In their preambles the two CEOS agreements identify common ground that provides the context for the principles which are discussed later. The preambles note the following common issues which are important for the development of Earth observation data policy:

- There is a wide range of uses of Earth observation data, including global change, climate change, environmental research, environmental monitoring, commercial applications, and operational observations such as those in the proposed global observing systems for the climate, the oceans and the land.
- Earth observation data are essential for some operational environmental applications.
- Earth observation data have potential economic and social benefits.
- Both data suppliers and users should respect the investment by governments in Earth observation systems, and acknowledge the value which this investment has provided.
- Different organisations within CEOS have different policy aims, e.g. maximising use or shifting funding responsibility to users.
- There are legal regimes in existence which guide some of the conditions of data provision and access.
- Members of CEOS have acknowledged the importance of a long-term commitment to data supply.
- CEOS members recognise a common goal of providing data to global change researchers from all missions on a consistent basis reflecting primarily the cost of filling the user request.

While these preambles are not the agreements themselves, they do provide important perspectives on Earth observation data policy. In particular, there is a recognition that: (1) Earth observation data have value and this value is a result of government investment; (2) global change researchers should be in a special category for pricing policy; and (3) there is a commitment to long-term Earth observation data supply, albeit to fulfil different policy objectives.

CEOS Principles

Having provided the context, this section now turns to a discussion of the CEOS principles themselves. The section is organised by presenting the comparable CEOS statements first, followed by the implications for Earth observation data policy. The principles are prefaced by their origin within CEOS, namely either those in support of global change research (GCR) or those in support of operational environmental use for the public benefit (OP).

Archives   *GCR. Preservation of all data needed for long-term global change/climate and environmental research and monitoring is required.*

Typically, Earth observation missions have funding for archiving their data for up to 10 years after the completion of the mission. Because of the potential value of the data, especially for global and climate change studies, there will need to be a provision for long-term archiving of Earth observation data beyond 10 years.

Archiving technology has made substantial strides in the last five years, and it is now possible to record large volumes of data on relatively small physical media. What is still required is a resolution on which organisations should take long-term responsibility for Earth observation data. Who should be responsible for safeguarding these valuable data for the next 100 or 200 years when global change researchers will need access to long runs of data?

Metadata   *GCR. Data archives should include easily accessible information about the data holdings, including quality assessments, supporting ancillary information, and guidance and aids for locating the data.*

OP. CEOS data suppliers should provide easily accessible information about the data and related mission parameters, including quality assessments, supporting ancillary information, and guidance and aids for locating and obtaining the data.

Metadata are those data which describe the characteristics of a data set. Accurate and up-to-date metadata that correspond exactly with their accompanying data are very important for the exploitation of Earth observation data. Even when the data themselves are poor representations of the environment, because of sensor saturation for example, the metadata must be accurate so that the quality of the data can be reliably determined. Accurate

Existing policies and policy making processes

metadata are necessary to cultivate confidence in the Earth observation data: if metadata are misleading or incorrect then confidence in using Earth observation data will diminish.

Metadata are, in essence, part of a data set or product. Calibration and validation data are essential for the correct analysis of Earth observation data. For example, SPOT products at levels 1A or 1B include the values of the calibration coefficients from the CCDs on the SPOT HRV instrument, together with the algorithms used to correct the raw data.

Standards *GCR. International standards – including those generated by the CEOS Working Group on Data – should be used to the greatest extent possible for recording/storage media and for processing and communication of data sets.*

OP. Recognised standards – to be defined and developed in common, including those generated by CEOS working groups – should be used to the greatest extent practical for recording/storage media and for the processing and communication of data sets.

These principles call for the use of international or recognised standards *to the greatest extent possible*. There are existing international standards for environmental data, and the World Meteorological Organisation standards are the most widely used environmental information exchange standards in practice. There are, however, problems of principle and practice. While there is a consensus that there are benefits from using standards, there are practical difficulties in full and complete implementation due to factors such as differences in the computer systems used. For example, all 14 Landsat receiving stations around the world implement data storage formats in a different manner. At receiving stations in countries with an acute lack of computer compatible tapes (CCTs), the data recording formats are designed to minimise the consumption of CCTs rather than to maximise the ease of user access to the data held on the CCTs.

This points to a weakness of CEOS: it is an international club working on the basis of best efforts between partners, yet it is not in a position to do anything other than encourage itself, i.e. its members, to follow its own principles. Not all of its members follow all the CEOS principles.

Sharing *GCR. Maximising the use of satellite data is a fundamental objective. An exchange/sharing mechanism among CEOS members is an essential first step to maximising use.*

OP. To optimise the use of data for operational environmental use for the public benefit, CEOS members should establish appropriate data provision mechanisms.

OP. Real-time and/or archived data for operational environmental use for the public benefit should be made available on time scales compatible with user requirements and within agency capabilities.

These CEOS principles recognise the need to provide and share data in line with user requirements, particularly in relation to time scales. If suitable standards are developed then data exchange/sharing should present few problems. There may be some questions over provision of near real time data, although the principles accept that data can only be made available in line with agency capabilities.

Data sharing is in principle beneficial. A potential danger can arise when sharing data crosses the boundaries between different categories of use of data and leads to the violation of licence agreements.

Access   *GCR. Non-discriminatory access to satellite data by non-CEOS members for global change/climate and environmental research and monitoring is essential. This should be achieved within the framework of the exchange and sharing mechanisms set up by CEOS members.*

When combined with the preamble point that there is a common goal of providing data to global change researchers at a marginal cost price, this principle has serious implications for Earth observation data pricing policy. This could allow a large category of users from all countries of the world access to data at marginal cost prices. The arguments concerning this issue are presented in the discussion of pricing policy in Chapter 6.

Exclusive use period   *GCR. Programmes should have no exclusive period of data use. Where the need to provide validated data is recognised, any initial period of exclusive use should be limited and explicitly defined. The goal should be release of data in some preliminary form within three months after the start of routine data acquisition.*

OP. Programs should have no exclusive period of data use except where there is a need to provide for data validation. An initial period of exclusive data use should be limited and explicitly defined. The goal should be release of data in some preliminary form within three months after the start of routine data acquisition.

This principle encourages open access to data as soon after the launch of an Earth observation satellite as is possible, and certainly within three months of routine data capture. Sadly this principle has not always been followed. In the past some NASA Earth observation data have never been openly released and have always remained within the domain of the Principal Investigator. User communities have also expressed concern over access to certain of the data from ESA's ERS satellites because of over-zealous Principal Investigators.

Harmonisation   *GCR. Criteria and priorities for data acquisition, archiving, and purging should be harmonised.*

OP. Criteria and priorities for data acquisition, processing, distribution, preservation, archiving and purging should be harmonised to take into account the needs of users of data for operational environmental use for the public benefit.

CEOS recognises the importance of collaborating on the ways in which Earth observation data are acquired, stored and (if necessary) purged. These principles concern the criteria and the priorities rather than the ensuing actions themselves. However, it does mean that Earth observation data providers should take explicit account of the plans of other organisations when planning their own data acquisition and safeguarding strategy.

2.3 International Earth Observing System

Participation

CEOS has a wide membership which includes both the suppliers of data and the organisations involved in the use of the data, such as the European Commission and the International Geosphere Biosphere Programme. A smaller group of agencies has also been formed which comprises the main public sector suppliers of Earth observation data. These are the agencies of the International Earth Observing System (IEOS) and the group consists of the relevant supplier organisations of Europe, Japan and the United States which planned the original Polar Platforms.

The IEOS agencies have produced a set of principles to guide their collaboration on data access and utilisation. In developing the IEOS principles, which are detailed below, the agencies envisaged the implementation of the principles to begin with the new Earth observation initiatives during the late 1990s (ESA 1994d), namely:

- Envisat-1 for ESA
- EOS-AM1 for NASA, as the start of the US Earth Observing System
- NOAA-N for NOAA, as the start of the NOAA Polar Orbiting Operational Environmental Satellites missions
- ADEOS for NASDA, as the start of the Japanese Earth Observing System
- TRMM for NASDA/NASA

IEOS Principles

1. All IEOS data will be available for peaceful purposes to all users on a non-discriminatory basis and in a timely manner.
2. There will be no period of exclusive data use. Where the need to provide validated data is recognised, any initial period of exclusive data use should be limited and explicitly defined. The goal should be release of data in some preliminary form within three months after the start of routine reception of instrument data.

3. All IEOS data will be available for the use of each of the agencies and its designated users at the lowest possible cost for non-commercial use in the following categories: research, applications and operational use for the public benefit.
4. Agencies which designate users for research use and for application use will do so through an Announcement of Opportunity or similar process. The designation will include a definition of the data to be provided. Research users shall be required to submit their results for publication in the scientific literature and applications users shall be required to publish their results in a technical report; both shall be required to provide their results to the designating agency and to the data-providing agency.
5. Any of the agencies may designate national users of the respective countries or member states of the agencies as it deems appropriate to be given access to all IEOS data at the lowest possible cost for non-commercial operational use for the public benefit, provided the designating agency assumes responsibility for ensuring that all the terms and conditions for data use are met. This use will have to be reported to the data-providing agency on the basis of commonly agreed criteria including type, usage and final destination of the data. Designation of users outside the national territory of the agencies or their member states (e.g. international organisations and agencies in non-participating countries) for non-commercial operational use for the public benefit will be done only with the agreement of the data-providing agency.
6. For purposes other than 3 above, the specified data will be made available in accordance with terms and conditions to be established by the data-providing agency.
7. Each data-providing agency will fulfil the data requests of the other agencies and their designated users to the maximum extent possible. In the event that these data requests exceed the data-providing agency's capacity, the data-providing agency and the designating agency will pursue alternative arrangements to fulfil such requests.
8. All data required by the agencies and their designated users will be made available on condition that the recipient agrees to applicable intellectual property rights, terms and conditions and/or proprietary rights consistent with these data exchange principles, and ensures that the data shall not be distributed to non-designated parties, nor used in ways other than those for which the data were provided, without the written consent of the data-providing agency.
9. Any of the agencies may delegate some of its functions to other entities; in which case, such agency will remain responsible for ensuring compliance with these data exchange principles.
10. Agencies will harmonise criteria and priorities for data acquisition,

Existing policies and policy making processes

archiving and purging in consultation with other relevant organisations.

Definitions

These IEOS principles use many terms whose interpretation can give rise to sensitivities. As a result, the principles carry with them the following operational definitions of some key terms.

- *Applications use* of data is a limited proof of concept study.
- *Data* means original Earth observation sensor output and higher level products created from it by the data-providing agency as part of the standard set of products.
- *Lowest possible cost* for designated users is a cost level which is no more than the additional cost of resources required to fulfil a specific user request.
- *Non-commercial use* is the use of Earth observation data to provide a service for the public benefit as distinguished from conferring an economic advantage on a particular user or group of users.
- *Non-commercial operational use for the public benefit* is the use of Earth observation data to provide a regular service for the public benefit as distinguished from conferring an economic advantage on a particular user or group of users.
- *Non-discriminatory basis* means that all users in a clearly defined data use category can obtain data on the same terms and conditions.
- *Research use* is within a study or investigation that aims to establish facts or principles.

Advantages and Disadvantages

The IEOS principles have not been completely agreed by all the partners. It seems likely that they will be interpreted as general principles to guide bilateral collaboration between data-providing agencies rather than as a set of binding statements across all IEOS organisations. For example, ESA has accepted the IEOS principles at the level of the Programme Board for Earth Observation, but to be formally binding on ESA they would require unanimous approval by the ESA Council.

The IEOS principles do give advantages to the data-providing agencies. Because of data sharing opportunities, each agency will have access to a suite of data which is much more extensive than that agency could provide by itself. This means in turn that each agency's users will be in a better position to gain access to data for research, applications and operational use in a public-good sense. This will give greater stability to the development of the user community because users will potentially have a wider range of access to Earth observation data.

On the other hand, there are disadvantages in full agreement to the IEOS principles. The principles suggest a stronger form of agreement among participating agencies than do the CEOS principles. Two themes in particular stand out. First, the IEOS principles have a clear price commitment. For users who are using Earth observation data in a public-good sense and are not using the data to derive a profit, there is a clear intention that those users should be able to acquire the data at the lowest possible cost. This cost means the extra, marginal cost beyond the space and ground infrastructure that must already be in place and funded by the data-providing agency. Therefore, users in this category do not contribute financially to the core space and ground infrastructure.

Second, the extent of the data flows between agencies and their designated users can be large because each data-providing agency will fulfil the data requests of the other agencies and their designated users to the maximum extent possible. The term 'maximum extent possible' is subsequently refined in the principles to mean the extent possible within available resources, but the starting point envisages the potential for extensive data exchange between agencies. Agencies will have to consider further investments to acquire data for their users or to respond to requests from other data-providing agencies. When the data are made available to users through other data-providing agencies there is an open question on which agency will monitor the correct application of the IEOS principles to constrain use to the defined categories of data use.

2.4 International Geosphere Biosphere Programme

As noted earlier, the International Geosphere Biosphere Programme is one of the programmes of the International Council of Scientific Unions. In this context it is important for the IGBP to have an agreed data policy for all the data, including Earth observation data, that are used in its core projects and related activities.

Using the data principles of ICSU and the World Data Center (WDC) system, the IGBP's Scientific Committee Meeting accepted the following data policy principles in December 1994 (Townshend 1996).

1. The IGBP requires an early and continuing commitment to the establishment, maintenance, validation, description, accessibility and distribution of high-quality, long-term data sets.
2. Full and open sharing of the full suite of global data sets for all global change researchers is a fundamental objective.
3. Preservation of all data needed for long-term global change research is

Existing policies and policy making processes

required. For each and every global change data parameter, there should be one explicitly designated archive. Procedures and criteria for setting priorities for data acquisition, retention and purging should be developed by participating agencies, both nationally and internationally. A clearing-house process should be established to prevent the purging and loss of important data sets.
4. Data archives must include easily accessible information about the data holdings, including quality assessments, supporting ancillary information, and guidance and aids for locating and obtaining the data.
5. International and where appropriate suitable national standards should be used to the greatest extent possible for media and for processing and communication of global data sets.
6. Data should be provided at the lowest possible cost to global change researchers in the interest of full and open access to data. This cost should, as a first principle, be no more than the marginal cost of filling a specific user request. Agencies should act to streamline administrative arrangements for exchanging data among researchers.
7. For those programmes in which selected principal investigators have initial periods of exclusive data use, data should be made openly available as soon as they become widely useful. In each case the funding agency should explicitly define the duration of any exclusive use period.

The IGBP principles are clearly in harmony with the IEOS principles and the CEOS principles, particularly over accessibility and data prices. The IGBP principles, as will be seen later in this chapter, are in even closer harmony with the US principles on data for global change research.

While the principles have been agreed, there is an open question on the implementation of the principles. In listing the principles, Townshend (1996) notes that IGBP is considering 10 issues that are significant to the full implementation of the principles. Of high priority in these 10 issues are the cost implications of making data available and the responsibilities for the long-term preservation of data:

It would be prohibitively costly for any (IGBP) Core Project or Framework Activity to make literally all of its data available.

How will long term archiving be achieved? This includes not merely technical issues but also those concerned with how responsibility is assumed and maintained.

So while full and open sharing of contemporary and archived Earth observation data in IGBP is both sensible and desirable, there may be practical limitations, largely driven by cost, on the ways and the extent to which the principles can be implemented.

3 REGIONAL

3.1 Introduction

Having discussed Earth observation data policy statements at the level of global organisations, this section examines the data policies of the main regional player, namely Europe. The Earth observation data policy statements of ESA, EUMETSAT and the European Science Foundation are examined in this section.

3.2 European Space Agency

Overall Principles

The European Space Agency was established in its present form in 1975. ESA has the 14 member states shown in Figure 3.1 plus Canada, which has a special agreement to participate in ESA programmes. ESA's purpose is to provide for and to promote, for exclusively peaceful purposes, cooperation among European states in space research and technology and their applications.

ESA has agreed a set of rules concerning information and data (ESA 1989) which apply to all its activities. The rules describe the conditions of ownership, intellectual property and dissemination of data and information for the following categories of ESA activities:

- data and information created by ESA staff members and fellows;
- data and information created by ESA contractors;
- data and information relating to payloads flown on space vehicles on which the Agency provides flight opportunities;
- exchange and dissemination of data and information between ESA member states and with non-member states.

In defining its rules, ESA (1989) explains why the rules have been established:

This policy and practice is meant to contribute to the carrying out of the Agency's mission and, in particular, to raise the level of European scientific research, improve the world competitiveness of European industry, ensure sound relations between the Agency and those making use of flight opportunities offered by it and promote the adoption by the private sector of products and services developed by the Agency.

In principle, these rules should flow down via the ESA programme boards to find implementation in data policies and in detailed implementation within the individual space missions.

Existing policies and policy making processes

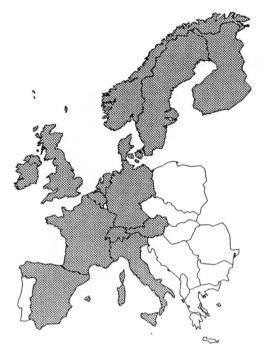

Figure 3.1 European Member States of the European Space Agency

The ERS Missions

ESA has launched two polar orbiting Earth observation satellites, ERS-1 in 1991 and ERS-2 in 1995. The satellites are owned and operated by ESA on behalf of the 13 member states participating in the ERS programme.

As noted in Chapter 2, the ESA member states have agreed a data policy to cover the data from the ERS satellites. The purpose of the data policy (ESA 1994c) is as follows:

> to promote the scientific/technological (research) application and commercial objectives of the programme in a balanced way, whilst stimulating the broadest possible involvement of the Earth science community and at the same time paving the way to an operational ocean/ice space remote sensing service with broad involvement of application users in the public and private sector.

The data policy document from which this statement of purpose is taken addresses a wide range of topics, including legal protection of data, licences for access to the data, pricing policy, pricing structures, special access conditions for particular groups of users and arrangements for data distribution and archiving.

While ERS data are made available in an open and non-discriminatory way, in line with the UN Principles on Remote Sensing, ESA retains full title and ownership rights to the data. Use of the data is then subject to *licence to use* agreements which control the distribution of the data and which vary with ERS products and the territory of distribution.

The overall policy on pricing is that prices should evolve in a phased way to reflect the progressive acceptance of the value of the data for applications and/or commercial purposes. Special access conditions are provided to particular groups of users cooperating with ESA. Table 3.1 shows the user categories that are recognised in the ERS data policy and the pricing policy conditions that attach to each category.

The pricing policy is implemented through the commercial distributor of ERS data, the ERS Consortium, which comprises Eurimage, Radarsat International and SPOT Image. As the implementation of the pricing policy for different users in Table 3.1, the 1996 price list for ERS data (Eurimage 1996) distinguishes two main price categories: a full price (the price for commercial users in Table 3.1), and a price for research users (research and demonstra-

Table 3.1 Categories of Users of ERS data. Source: ESA (1994c)

Category of user	Pricing policy
A. Internal use, comprising calibration and validation, principal investigator use, provision of data to Announcement of Opportunity instrument providers, and organisations cooperating in pilot and demonstration projects	Data are provided free of charge, with a proviso that they are only used for the approved purpose and are not made available to third parties
B. Research and demonstration use for the development of innovative techniques in approved projects	Data price is the cost of fulfilling the user request, normally known as the marginal cost
C. National meteorological services belonging to WMO member states, plus ECMWF and EUMETSAT	Low bit rate (LBR) fast delivery (FD) data free of charge for authorised uses
D. Operational services for public utility	Fast delivery (LBR and SAR) and low resolution SAR products at marginal cost for authorised uses
E. Commercial users	A complicated set of arrangements which in practice imposes a notional price and not the full economic cost of capturing the data and producing the products

tion use plus operational services for public utility in Table 3.1). Table 3.2 gives examples of the prices charged by Eurimage for some of the data products.

The price policy and the price list include an extra royalty of approximately 20% of the full price to be paid by buyers in countries that are not part of the ERS programme.

Envisat

ESA is now developing its Envisat mission as part of the Polar Orbiting Earth observation Mission (POEM-1) programme (ESA 1994a). In the declaration on the POEM-1 programme, ESA member states have agreed to define a data policy for Envisat:

[The Participating States] . . . AGREE to define in due course an Agency data distribution policy for each of the missions (that is Envisat-1 and Metop-1), also taking into account the experience accrued on the ERS data distribution, consistent with commitments entered into by the Agency in relevant agreements and towards instrument providers and with a view to securing data continuity to the users' community.

This statement is wide ranging in its objectives, and emphasises the lessons learned from the experience with ERS, ESA's international agreements such as those of CEOS, and importantly the long-term continuity of data supply. These have led ESA to identify explicitly the following three categories of use of Envisat data (ESA 1996a):

- the scientific domain, including all areas of research in Earth sciences which are addressed by the Envisat payload (e.g. oceanography, global warming, climate change, ozone depletion);

Table 3.2 Prices for a Sample of ERS Data Products. Source: Eurimage (1996)

Product type	Full price (ECU)	Research user price (ECU)
SAR fast delivery image	500	200
SAR precision image	1200	300
SAR terrain geocoded image	2300	1000
SAR wave mode on-line fast delivery per month	400	200
Wind scatterometer on-line fast delivery per month	350	175
Radar altimeter on-line fast delivery per month	350	175
ATSR sea surface temperature	80	40

Note: 1 ECU = approximately US$1.3.

- the operational and public benefit domain, in particular in support of operational organisations (e.g. meteorology, environmental monitoring) working within participating states (of the Envisat programme) or within international entities recognised by member states;
- the commercial domain, including value added companies, commercial distributors and operators, application developers, service providers and end-users.

ESA explicitly recognises the need for action to secure data continuity in the long term by paving the way for future operational missions by including in its options for Envisat data policy three new issues in its policy development:

- the improvement of the level of use of data from Earth observation satellites in operational and commercial activities;
- the improvement of the economic return either directly to ESA or to appropriate entities from the exploitation of the ESA satellites;
- and significantly, the preparation of more advanced financing schemes for ESA Earth observation satellites, in particular by ensuring participation, in both the development and the operation phases, of important user entities.

These issues emphasise the need for more consideration of the higher end of the value chain of Earth observation, and point to a fuller consideration of Earth observation from a research and development activity into an operational activity.

3.3 EUMETSAT

The European Organisation for the Exploitation of Meteorological Satellites (EUMETSAT) is an international organisation funded through its 17 member states and their National Meteorological Services (NMSs). The prime objective of EUMETSAT is to obtain meteorological data by satellite on behalf of its members. These data have so far been provided by the geostationary satellites of the Meteosat series, but EUMETSAT also has plans to launch the EUMETSAT Polar System in the 21st century.

In June 1991 the EUMETSAT Council approved a document describing EUMETSAT's current policy on access to its data. The policy is founded on the consideration that the data concerned represent a special resource owned by the NMSs of the EUMETSAT member states. The policy document itself contains a set of principles on data distribution and charging agreed by the Council in 1990.

The policy states that EUMETSAT has full ownership and all utilisation rights of the Meteosat satellites and their data. Data, products and services are put at the disposal of the National Meteorological Services of the EUMETSAT member states at no further cost, and the NMSs are responsible for the distribution and charging of data within their own territory.

Existing policies and policy making processes   61

In 1994 the EUMETSAT Council reached agreement on a more active policy to protect access to EUMETSAT data by encrypting certain of the data transmissions. The subject of encryption of EUMETSAT data is discussed in Chapter 5.

3.4 European Science Foundation

The European Science Foundation (ESF) has taken action to coordinate the views of the European scientific community on the future development of Earth observation in Europe because it recognises that in the past this community has had no channel 'for articulating its ideas, defining its priorities and offering its advice in the form of a coherent European scientific consensus view' (ESF 1992).

To address this need the European Earth Observation Panel (EEOP) was created by the ESF as a panel of its European Space Science Committee. The EEOP convened two workshops in 1992 which resulted in the preparation of a strategy paper entitled 'A Strategy for Earth Observation from Space', which presents a view on data policy from the European scientific community. The ESF paper makes the following key points on Earth observation data policy:

- It is necessary to view Earth observation as an end to end system in which adequate resources are provided not only to fund the space and ground infrastructures, but also to enable the final assimilation of Earth observation data into models that can provide an integrated description of the complete environmental system.
- A proper balance of resource investment between the space and ground segments of the Earth observing system is required.
- Earth observation data should be provided at low cost (free or marginal cost) to stimulate both the scientific understanding of the Earth system and the development of operational and commercial applications. The primary goal should be to encourage maximum use of the data for the objectives for which the space and ground segments were established.
- International collaboration and independent scientific review of programmes should be intensified.

4 NATIONAL

4.1 Introduction

Having discussed global and European Earth observation data policies, this section gives examples of three nations that have developed Earth observation data policies, namely France, the UK and the USA. Most nations active in

Earth observation have produced some statements on Earth observation data policy, e.g. Australia, Brazil and Germany (Harris and Krawec 1993a), but the three chosen here illustrate particularly well the main Earth observation data policy issues involved at a national level.

4.2 France

In France an interministerial group has developed a policy document on Earth observation data dissemination from space (Synthesis 1995). The interministerial group consisted of 17 sections of French government departments and related organisations, including the French meteorological service, the space agency, the ministry of the environment and the national mapping agency. France is particularly significant in this regard because of its investment in the SPOT programme and its highly active participation in ESA and EUMETSAT.

The policy document sets the context for the development of a French position by discussion of the challenges facing Earth observation, the legal framework in which Earth observation is operating and a summary of Earth observation data policy developments in the USA, the European Union, the World Meteorological Organization, ESA, EUMETSAT and CEOS.

Having reviewed the context, the policy document identifies the basic principle of the French position: to achieve a return on the state's investment in satellite Earth observation programmes. The document recommends the following four components to this basic principle:

- Because of the major efforts granted by the state toward the development of Earth observation systems, and to guarantee the continuation of such systems, the basic principle is established that the dissemination of Earth observation data will lead to a return on investment.
- The nature of this return on investment may be scientific, humanitarian, strategic, operational or financial (see also Boutros-Ghali 1994).
- The principle of return on investment is implemented in forms defined according to the objectives of the considered space programme and the mission of the organisation in charge of analysing and disseminating the data.
- This principle applies to all of the French (state) organisations. In particular, public organisations have to assess the returns on investment within the framework of the provisions governing their status.

The principle of a return on investment is clearly stated. However, it is not narrowly interpreted in a financial sense; the return is related to the objectives of the programme that require the investment. This means that a greater recognition is required of how the return on investment will be measured. For programmes that have a commercial component to their objectives (e.g. ESA's

Envisat-1 mission), it is imperative that the parent organisation identifies what measures it will take to gauge the success of the return on investment.

The French policy document suggests three types of satellite Earth observation programme, each of which has different conditions for its return on investment. The three types are shown in Table 3.3 along with a summary of the type of return on investment.

4.3 United Kingdom

In 1990 the UK established the Inter-Agency Committee on Global Environmental Change (IACGEC). An advisory group was established, led by the British National Space Centre, to advise on the issues of management of data relevant to global change, including Earth observation data. The participation in the advisory group included UK research councils (NERC 1996) and government departments, and its remit was to examine the major prevailing policy and technical issues in international data handling.

The report of the advisory group (IACGEC 1992) made a set of recommendations as the basis for an integrated national and international approach to data management for global environmental change. The main elements of the recommendations are listed below.

- Earth observation data and other global environmental change data should be held in accessible and usable forms. International activities to improve metadata provision, quality assurance and the development of standards should be promoted to ensure greater accessibility of data.
- The IACGEC agencies should recognise that there is no such thing as free data. Somewhere somebody pays. Finances should be explicit and trans-

Table 3.3 Three Types of Earth Observation Programme and the Expected Return on Investment. Source: Synthesis (1995)

Type of Earth observation programme	Return on investment
Strictly experimental programmes intended for scientific research	Progress in facts, knowledge, principles and techniques
Pre-operational or operational programmes for the public benefit, in particular weather forecasting, environmental monitoring, climate monitoring and natural risk assessment	Exchange or pooling of data, knowledge, derived information and information services for the public benefit
Pre-operational or operational Earth observation programmes within which the data are to be disseminated on the basis of a financial return	The return on investment is financial

parent so that when decisions are made on reducing costs of data access for scientific research then those decisions are informed on the financial implications.
- International exchange of data is applauded so that global change researchers can gain access to more environmental data than is possible with solely UK-supported missions.
- The value of global change data should be recognised and monitored so that, regardless of whether individual global change researchers pay for access to the data, the value can be identified to all interested organisations.

4.4 United States

The US Global Change Research Program was established by President Reagan and included as a Presidential Initiative by President Bush in the 1990 financial year (FY) US budget. It subsequently became formalised in the Global Change Research Act of 1990 which aims to understand, assess, predict and respond to global change arising through both human-induced and natural processes, while at the same time improving the effectiveness and productivity of US efforts in global change research (NSTC 1996).

The recent annual budgets for the USGCRP have been US$1.72 billion in 1995 and US$1.61 billion in 1996. The planned budget for 1997 is US$1.73 billion. Twelve US government agencies are involved in funding the USGCRP, including the departments of agriculture, commerce, defence, energy and transportation and the National Science Foundation.

The largest spender within the USGCRP is NASA, with over 70% of the USGCRP budget in recent years. The NASA component of the USGCRP is for the Mission to Planet Earth (CES 1995), i.e. the current and future Earth observation systems including EOS, Landsat, TOPEX/POSEIDON and TRMM. For 1997 the financial authority for NASA's Mission to Planet Earth is US$1.33 billion (NASA 1996).

The USGCRP has agreed the following set of policy statements on data management for global change research to apply to all the relevant data in the USGCRP, including Earth observation data.

- The US Global Change Research Program requires an early and continuing commitment to the establishment, maintenance, validation, description, accessibility and distribution of high-quality, long-term data sets.
- Full and open sharing of the full suite of global data sets for all global change researchers is a fundamental objective.
- Preservation of all data needed for long-term global change research is required. For each and every global change data parameter, there should be at least one explicitly designated archive. Procedures and criteria for

Existing policies and policy making processes 65

setting priorities for data acquisition, retention and purging should be developed by participating agencies, both nationally and internationally. A clearing-house process should be established to prevent the purging and loss of important data sets.
- Data archives must include easily accessible information about the data holdings, including quality assessments, supporting ancillary information, and guidance and aids for locating and obtaining the data.
- National and international standards should be used to the greatest extent possible for media and for processing and communication of global data sets.
- Data should be provided at the lowest possible cost to global change researchers in the interest of full and open access to data. This cost should, as a first principle, be no more than the marginal cost of filling a specific user request. Agencies should act to streamline administrative arrangements for exchanging data among researchers.
- For those programmes in which selected principal investigators have initial periods of exclusive data use, data should be made openly available as soon as they become widely useful. In each case the funding agency should explicitly define the duration of any exclusive use period.

These policy statements are almost identical to those of the IGBP described earlier in this chapter, illustrating the heritage of the IGBP statements and the enthusiastic part played by participants of the USGCRP in Earth observation and global change research worldwide.

The US policy statements on data management for global change research together with the funding levels of US$1.7 billion per annum clearly illustrate the US commitment to making Earth observation data widely available for scientific research purposes. To back up this policy, NASA is spending approximately a quarter of a billion dollars per annum (NSTC 1996) on building the EOS Data and Information System (EOSDIS), which will give users very high quality access to the data from Mission to Planet Earth and to data from other spaceborne Earth observation systems.

5 CONCLUSION

The role of the user is being enhanced in the development of Earth observation data policy. There is now participation of user community organisations in the Committee on Earth Observation satellites, although not yet in the IEOS.

The international scientific research actions are developing policy statements and the global monitoring programmes have developed a common space data plan. This is important because the supplier organisations can respond more readily to coherent international agreements than to diverse individual statements.

The science community is making its voice heard more clearly through routes such as IGBP, the European Science Foundation and national research councils. This highlights the relatively poor representation on international bodies of the commercial sector, which is a strong player at the level of individual space agencies, but not so strong at the global level.

At the same time as users are having a stronger voice in Earth observation data policy, the supplier organisations are improving the clarity and precision of their statements about their own data policies. Improved definitions of objectives should lead to better international communications and less conflict over Earth observation data policy.

4
Physical Access to Earth Observation Data

1 INTRODUCTION

Having explored the factors that influence Earth observation data policy and the existing data policies, this chapter now turns to the issues of the physical access to Earth observation data. The purpose of the chapter is to review the factors that affect access to Earth observation data by the eventual users. This is fundamentally about control of the data chain and the pressures which act upon that chain. The physical access to Earth observation data is important to all users because the various steps condition how Earth observation data are made available to users.

The chapter opens by discussing the methods of acquisition of Earth observation data at ground stations, and then goes on to examine ground segment issues and distribution of data.

2 DIRECT ACQUISITION

2.1 Low Data Rate

The simplest access to Earth observation data is from the meteorological satellites that transmit data at a low data rate in real time. Meteosat, GOES, GMS and NOAA AVHRR data can all be collected by a small receiving antenna connected to a PC equipped with suitable data handling software. The antenna is typically only 1 m in diameter and the data rate is approximately 1 kbit s^{-1}.

Data reception is possible when the satellite is in line of sight of the receiving station. For the geostationary satellites this means being within the zone of the sensed disc which makes up the satellite image: this is about 55° in all directions from the sub-satellite point. For the polar orbiting satellites the coverage is determined by the amount of time the satellite emerges from below the horizon, traverses the sky above the receiving station and descends again beyond the horizon. Depending upon the altitude of the satellite, this is normally a circular area about 2600 km radius centred on the receiving station.

The costs of building a suitable receiving station for these low rate data are low and can be of the order of a few thousand US dollars.

Most of the data are available with no conditions, except for some of the Meteosat data which are encrypted and the encryption key is available from EUMETSAT (see Chapter 5).

2.2 High Data Rate

Data from the high data rate instruments on board satellites such as Landsat, SPOT, ERS and ADEOS require much more sophisticated and costly receiving equipment. The X-band antenna required to receive these data is typically over 5 m in diameter and the data handling requires large-scale processing equipment. In addition, and of greater importance, the reception of these data requires the agreement of the satellite owner to capture the data through some form of licence to receive.

Data reception is only possible when the satellite is in direct line of sight of the ground station. As all the satellites are polar orbiting this means that data can be received when the satellite is within approximately 2600 km of the receiving station. Figure 4.1 shows the coverage of the stations that receive SPOT data. The greater coverage in polar regions is caused partly by the map projection used, and partly because the polar orbits of the satellites mean that there is a good capability to capture many overpasses at stations such as Kiruna in northern Sweden.

The cost of developing a receiving station capable of collecting data from a variety of Earth observation missions, such as Landsat, SPOT and ERS, is of the order of US$5 million. Such centres exist in many parts of the world, including Argentina, Thailand, Europe and the USA. From the overlapping coverages of the receiving stations, few parts of the land surface of the planet are omitted from the potential to collect Earth observation data of that area. For the regions that may not be included, mobile or transportable receiving stations have been developed. A transportable receiving station for ERS data reception developed by Matra Cap Systémes was sold to the US Department of Defense for a reported US$5 million and a transportable receiving station for the Antarctic has been deployed by Germany for ERS data reception.

The UK Overseas Development Administration has sponsored the development of a lower cost, transportable receiving station for use in developing countries to acquire local-coverage ERS SAR data. The receiving station can be low cost because some of the coverage and quality constraints can be relaxed; for example, the capability to acquire data when the satellite is

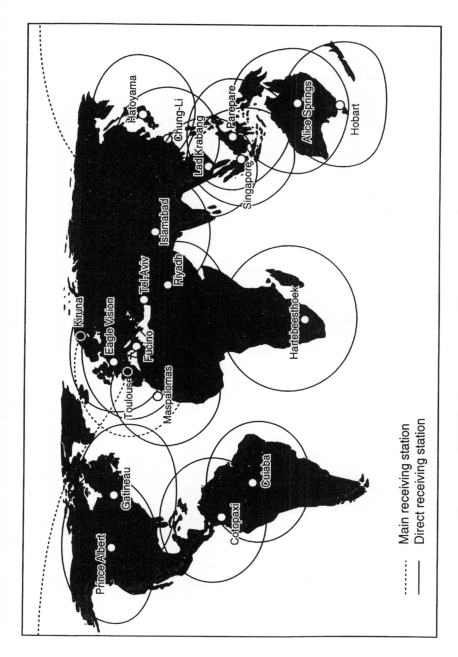

Figure 4.1 SPOT Receiving Stations and Their Coverage Zones

within 2° of the horizon can be relaxed to the acquisition of data only when the satellite is above 5° of the horizon.

Instruments such as the SAR on ERS and the TM on Landsat do produce high data rates themselves and require extensive processing. There are also instruments which produce low rate data but which still require sophisticated processing, e.g. the ATSR on ERS-1. Data from the lower data rate instruments are multiplexed with other data to make the downlink transmission efficient and so the low rate data are not easily accessible to the smaller receiving stations, which can acquire NOAA AVHRR or Meteosat data, because the data are embedded in other transmissions.

2.3 Alternative Data Routes

Tape Recorders

One alternative to direct broadcast transmissions is to use on-board tape recorders. These recorders are used to store data on-board the satellite until it is in line of sight of a suitable ground receiving station where the stored data can be broadcast to the ground.

ESA's Envisat-1 mission will have 14 or 15 orbits per day, depending on the orbit configuration (Logica 1996). The satellite will carry four tape recorders and each tape recorder will be capable of storing instrument data from one orbit of the Earth by the spacecraft. The receiving station at Kiruna in northern Sweden will be used to downlink the data either as it has just completed an orbit or at a later stage if the orbit is out of line of sight of Kiruna.

Data Relay

A second alternative to direct line of sight data reception is to use another satellite to provide communication between the Earth observation satellite and the ground station. A data relay satellite (DRS) has two main advantages for Earth observation missions. First, it can enable more flexibility in the control of the satellite and its payload, since the satellite is normally only visible for commanding for a few minutes per orbit. Second, a data relay satellite offers opportunities to collect global data and gives access to data from otherwise inaccessible parts of the world.

Landsat data have been collected this way by being routed via the Tracking and Data Relay Satellite System (TDRSS), and EOS will have this capability also. There are two TDRSS satellites in geostationary orbit on opposite sides of the globe, and it is therefore possible for a data transmission route to flow from Landsat in low Earth orbit to TDRSS to the ground receiving station.

For Envisat-1 there is a plan to use the Artemis data relay satellite. Artemis is scheduled to be launched by ESA into a geostationary orbit at 16.4° east longitude above the equator. It will carry a data relay payload to relay Envisat-1 data from the instruments on the spacecraft via Artemis to a User Earth Terminal in Europe.

With the Artemis DRS it will be possible for ground receiving stations in Europe to receive Envisat data direct as long as they have (1) Ka band reception capability, and (2) they fall within the specialised reception footprint of Artemis. The Artemis satellite will have a feeder link coverage to a large area of Europe. This coverage area is shown in Figure 4.2.

The use of a data relay satellite increases the capability of satellite owners and operators to acquire Earth observation data on a worldwide basis and so to circumvent local data receiving stations. This level of control is beneficial for rapid access to data of remote regions. Rapid access is valuable for hazard monitoring. Marine oil slicks can occur in remote locations, and it is helpful

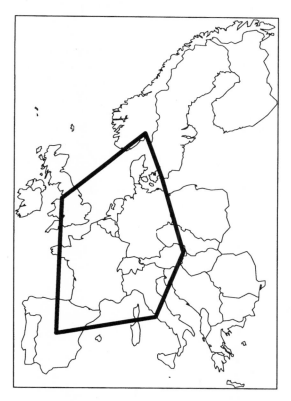

Figure 4.2 European Feeder Link Coverage from the Artemis Data Relay Satellite. Source: Logica (1996)

for environmental protection organisations to be able to monitor the slicks wherever they occur. Data relay satellites provide the means to collect Earth observation data of remote, inaccessible regions in near real time.

3 PROGRAMMING SATELLITE DATA ACQUISITIONS

3.1 ERS

Control of satellite data acquisition is an increasingly useful part of Earth observation. With the ERS missions, for example, the radars can only be switched on for seven minutes per orbit because of power supply constraints. Therefore it is necessary to program the ERS satellites so that they acquire the seven minutes' worth of data over the right geographical location(s).

Within ESA, two of the establishments (ESRIN and ESOC) combine together to prepare the data acquisition schedule for the ERS satellites. Figure 4.3 illustrates the process.

User requests are routed and collated via the ERS Consortium, which is the commercial agency for ERS data. The ERS Consortium then transmits the user requests to ESRIN at Frascati, near Rome. ESRIN prepares and validates the Preferred Exploitation Plan (PEP), which defines the way in which the ERS satellite should be commanded to acquire the correct data.

The PEP is transmitted to the ERS mission control centre at ESOC in Darmstadt, Germany. ESOC validates the PEP and then prepares a Detailed Mission Operations Plan (DMOP) which defines the actual scheduling of the data acquisitions. This DMOP is used to command the SAR acquisitions by the ERS satellite.

Users are typically required to request data acquisitions one month in

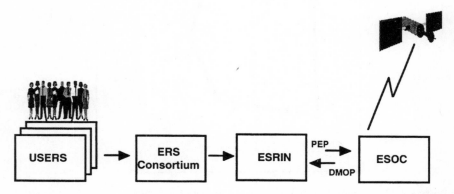

Figure 4.3 The Routeing for ERS-1 Data Acquisition Requests. Source: Logica (1996)

advance, but short-term requests, such as for monitoring environmental disasters, can also be accommodated. 'Short term' in this context means one day, for example with ERS SAR data collection following the *Sea Empress* accident off the coast of South Wales in 1996. Short term is becoming even shorter: Space Imaging offers a period of 90 minutes before satellite commanding to finalise the data acquisition plan. The plan is completed 10 minutes before the communication with the satellite can be established, and the data acquisition schedule is uploaded to the satellite in the first 90 seconds of the spacecraft overpass.

3.2 SPOT

The SPOT satellite is unusual in having a pointing capability for its instrument. This pointing capability is used to respond to user requests for data. The pointing capability is particularly useful in allowing the collection of data in cloudy conditions: the instrument can be pointed to avoid the cloudy areas.

SPOT Image has developed a red and blue programming scheme, which gives the opportunity for the SPOT satellite to be used flexibly to respond to a user request for data. This scheme is used, for example, in the European Union programme of Monitoring Agriculture by Remote Sensing (MARS) (Sharman 1995).

Red programming gives the users of SPOT data the highest priority of SPOT data acquisition and a contractual allocation of the satellite's resources. The SPOT satellite is programmed via the SPOT centre in Toulouse to point its instrument at the targets with the red programming requests, provided that there is no conflict in the pointing. *Blue programming* has a similar objective and operation, but blue programming requests have a lower priority than red programming requests and cost less to the user.

The costs of the red and blue programming add considerably to the data costs. SPOT scenes acquired by red programming cost approximately twice as much as archived SPOT scenes.

SPOT satellite programming is used in the monitoring of the potato crop in the UK. The Potato Marketing Board (PMB) has been using SPOT programming to acquire Earth observation data of the main areas growing potatoes. The PMB uses these data to check the returns from farmers: these returns give the geographical coordinates of the fields in which potatoes are being grown. Because of cloud cover problems it is not possible to guarantee SPOT data acquisition on specific days, so the PMB works with Logica UK Ltd and SPOT Image to program the SPOT satellite to acquire data of the main potato-growing regions within certain time periods (Pryor, Harris and Williams 1996).

Figure 4.4 shows the main potato-growing regions of Britain and the image

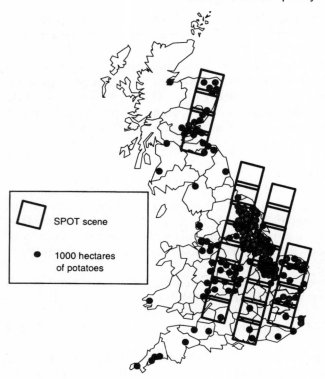

Figure 4.4 Orbit Tracks and Nominal Scenes of SPOT Data Acquired over the Main Potato-Growing Regions of Britain

areas of selected SPOT orbit tracks. Because the SPOT instrument can be pointed, the precise positioning of the track can be altered to acquire the data that are needed. The precise track and scene positions in Figure 4.4 are not definitive, although the alignment of the tracks is.

The capability to program the SPOT satellite allows the PMB to have a high probability of acquiring SPOT data of the regions of Britain growing potatoes within the growing season of May to September.

3.3 Space Imaging

Space Imaging Inc. has a new concept for access to Earth observation data. Space Imaging is developing a very high resolution satellite system with a smallest pixel size of 0.82 m. Access to Space Imaging data will be achieved either from a central receiving station in the USA, or by ground stations that are operated by regional affiliates. These regional affiliates will either buy an equity stake in Space Imaging or will pay an annual fee for data access. In return for this payment the regional affiliate will have the right to task the satellite, including a pointing capability, and to acquire the data directly at a regional ground station. In essence, this gives a regional affiliate dedicated control of and access to an Earth observation satellite for a given region, which as noted above is a circular zone of approximately 2600 km radius.

The initial investors in Space Imaging are Lockheed Martin (US$150 million), E-Systems (US$100 million) and Mitsubishi (US$25 million) (Anon 1995). Space Imaging has given indicative prices of US$25 million for a regional affiliate ground receiving station and US$25 million per annum for control of the satellite downlink to give the regional affiliate ownership of the data collected at the ground station.

Reception of Space Imaging data is ultimately dependent upon US government policy because under the terms of the licence granted to Space Imaging by the Department of Commerce in April 1994, the US government can request the satellite to be turned off during times when US national security is threatened.

4 GROUND SEGMENTS

4.1 Introduction

The next step in the discussion of access to Earth observation data is the ground segment. The discussion so far in this chapter has been about reception of data at an antenna on the ground. Once the data are received they must be processed to produce data and information products. This section describes the major ground segments that carry out that processing for SPOT, ERS, Envisat and the US Earth Observing System.

4.2 SPOT

The SPOT satellites have their own ground infrastructure system, consisting of a primary network and a secondary network. The primary network comprises two SRIS stations (Satellite Reception des Images Spatiales). One is located in Aussaguel, near Toulouse (SRIS-T), and the other at Kiruna, in northern Sweden (SRIS-K). These stations receive data transmitted directly by the SPOT satellite for Europe, North Africa and the north polar regions, together with data acquired over other areas and stored on the on-board tape recorders. Figure 4.5 shows the main elements of the ground segment.

Around the world there is a secondary network of receiving stations for SPOT data. Figure 4.1 shows the distribution of the SPOT Direct Receiving Stations (SDRS). Each SDRS carries out the following activities:

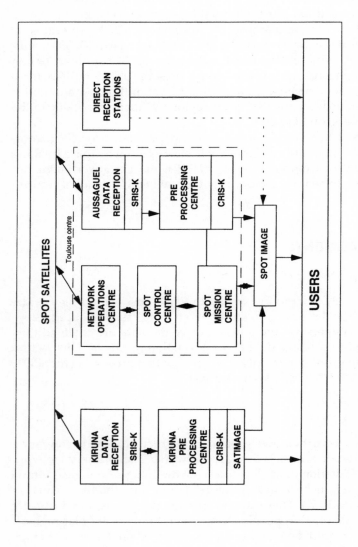

Figure 4.5 The SPOT Ground Segment. Source: Logica (1991)

- reception and archiving of image data relating to the station's coverage area;
- distribution of SPOT data in the country where the station is located, though not exclusively;
- transfer of data, where necessary, to SPOT Image in Toulouse for distribution to the rest of the world;
- supply of updates to the SPOT Image catalogue, so that the catalogue has metadata on all the SPOT data collected worldwide.

All SPOT data can be viewed as quick-look images through the on-line DALI system, which allows users to interrogate remotely the SPOT catalogue for information on all the SPOT data that has been acquired and is in the archive. Users can also gain access to the SPOT catalogue and some SPOT data at the SPOT WWW site:

http://www.spotimage.fr/

4.3 ERS

Data Reception

There are three downlinking components to ERS data transmission: reception by ESA stations; reception by ESA member states with their own receiving stations (called by ESA the national stations); and reception by the stations outside ESA member states (called 'the foreign stations' by ESA). The ERS data are then handled by the ERS ground segment shown in Figure 4.6.

The ERS ground segment can be divided into several main sections. The central data facility at ESRIN coordinates the data acquisitions (through ESOC) and the off-line data processing. It also communicates with the ERS Consortium, the commercial organisation that handles commercial user requests. The Processing and Archiving Facilities are responsible for the off-line processing of ERS data products. The sequencing of their processing is scheduled by the central data facility.

Rutherford Appleton Laboratory in the UK processes data from the ATSR instrument. Low bit rate data, such as wind scatterometer data, are transmitted via the meteorological Global Telecommunications Network, with nodes at the Italian and UK meteorological services. Finally, fast delivery SAR data are disseminated using the Broadband Data Dissemination Network, a part-time facility for rapid data transmission. Depending on the receiving station, there is also some data dissemination direct to users. In the UK, rapid access to SAR data can be achieved by running RAIDS (RApid Information Dissemination System) from the following WWW site:

http://www.demon.co.uk/raids/index.html

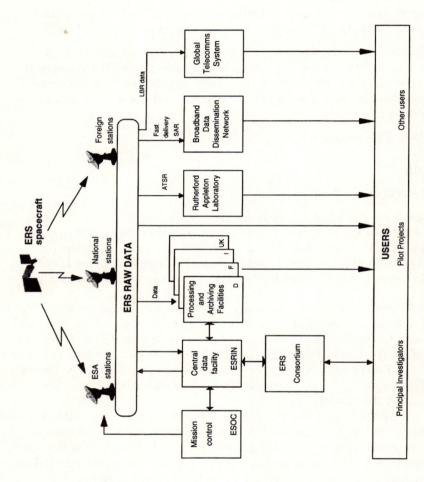

Figure 4.6 The ERS Ground Segment. Source: Logica (1993)

RAIDS is installed at West Freugh in Scotland, the site of the UK ERS receiving station, and provides electronic access to ERS SAR data received at West Freugh within 24 hours of data acquisition.

Processing and Archiving Facilities

ERS data processing is somewhat different in having geographically distributed processing that is under centralised control. The off-line processing of the data received from ESA's ERS missions is carried out at four Processing and Archiving Facilities (PAFs) in Europe. These four PAFs are at Brest, Farnborough, Oberpfaffenhofen and Matera. The systematic activities of the four PAFs are listed below. The activities are distributed according to the national interests of the PAFs.

- Brest, France

— Routine long-term archive of all global low bit rate (LBR) data, acquired by the ERS-1 stations at Kiruna, Maspalomas and Gatineau.
— Routine long-term secondary archive of global ATSR data.
— Routine data retrieval and generation of radar altimeter ocean products and supply of full data sets to the Oberpfaffenhofen PAF.
— Routine data retrieval and generation of ATSR microwave products.

- Farnborough, UK

— Routine long-term archive of SAR raw data acquired at the ERS-1 Kiruna station and of selected data sets from national and foreign stations acquired for ESA or where no long-term archiving is guaranteed.
— Routine long-term secondary archive of all global LBR data acquired by the ERS-1 stations at Kiruna, Maspalomas and Gatineau on optical discs generated at the Fucino LBR transcription facility.
— Routine data retrieval and generation of radar altimeter products – 100% of global data acquired over ice and land.
— Routine data retrieval and generation of ATSR products – 100% of global data acquired over ocean and ice.
— Routine primary archive of SAR processed data, ATSR raw and processed data, LBR data over ice and land and associated fast delivery products, ERS-1 calibration and validation data and products.

- Matera, Italy

— Routine long-term archive of all LBR real-time data over the Mediterranean area and of the SAR regional data sets acquired at the Fucino station and of selected SAR data sets from national and foreign stations acquired for ESA.

— Routine long-term archive of fast delivery products and their intermediate products processed at Fucino for regional purposes as well as the calibration and validation data for products generated at the Matera PAF.

- Oberpfaffenhofen, Germany
— Routine long-term archive of SAR raw data acquired over Antarctica and of selected SAR data sets from national and foreign stations acquired for ESA.
— Routine generation of higher level altimeter products and precision orbit calculations.

4.4 Envisat

ESA will launch its next Earth observation satellite, Envisat-1, at around the turn of the century. Envisat will both update and expand the data sets collected by the ERS satellites and will add new observation capability in atmospheric chemistry and marine biology. Envisat will carry the nine instruments shown in Table 4.1: five of these are instruments developed by ESA and four are developed and provided by individual ESA member states.

The ground segment for Envisat combines ESA facilities and national facilities in a data processing infrastructure to handle the data and produce a wide variety of products. The Envisat-1 ground segment concept document (ESA 1994e) lists 109 products to be produced routinely from the Envisat data: examples of some of these products are listed in Table 4.2.

ESA facilities will be used as part of the Envisat ground segment to plan and execute the operations of the satellite and to acquire the global data set, the regional data set of the European zone and any regional data for a part of the world covered by an ESA data relay satellite. The ESA facilities will be

Table 4.1 The Instrument Payload on Envisat-1. Source: ESA (1994e)

Instruments developed by ESA	Announcement of Opportunity instruments
- ASAR Advanced Synthetic Aperture Radar - GOMOS Global Ozone Monitoring by Occultation of Stars - MERIS Medium Resolution Imaging Spectrometer - MIPAS Michelson Interferometric Passive Atmospheric Sounder - RA-2 Advanced version of the ERS radar altimeter	- AATSR Advanced Along Track Scanning Radiometer - DORIS Doppler Orbitography and Radiopositioning Integrated by Satellite - SCARAB Scanner for Radiation Budget - SCIAMACHY Scanning Imaging Absorption Spectrometer for Atmospheric Chartography

Table 4.2 Example Envisat-1 Products. Source: ESA (1994e)

Instrument	Product	Systematic processing	On request processing	Time scale
AATSR	Sea surface temperature	✓		1 to 3 days
ASAR	Medium resolution radar image	✓		3 hours to 3 days
ASAR	Precision radar image		✓	3 hours to 2 weeks depending on processing centre
GOMOS	Atmospheric constituent profiles	✓		3 hours
MERIS	Raw spectrometer data	✓		1 day
MERIS	Coastal water concentrations		✓	1 day to 3 days

responsible for near real-time processing and dissemination to users of fast delivery products (typically disseminated within a few hours of data reception), and for the consolidation of much of the raw data from the Envisat mission.

The ESA facilities will be complemented by national facilities. These national facilities will include the provision of receiving stations in participating states and the provision of off-line processing and archiving facilities in Processing and Archiving Centres (PACs). The PACs are similar to the ERS PAFs, but with a name change for consistency of terminology in the Envisat ground segment.

Figure 4.7 gives an overview of the Envisat ground segment. The ground segment has three main sections: the ESA Payload Data Segment, the reception and processing by the participating states, and the external interfaces.

The ESA Payload Data Segment (PDS) consists of the Payload Data Control Centre which is located at ESRIN in Frascati, the Payload Data Handling Stations at Kiruna, Sweden and at ESRIN (with the associated data reception via DRS), and the Payload Data Acquisition Station at Fucino, near Rome. In addition, the ESA PDS has a Low rate Reference Archive Centre (LRAC) for archiving data from all the instruments except for ASAR and MERIS.

The PACs in the participating states will provide off-line precision processing of ASAR and MERIS data, the systematic production of geophysical products, archiving of fast delivery products and the distribution of off-line products to users.

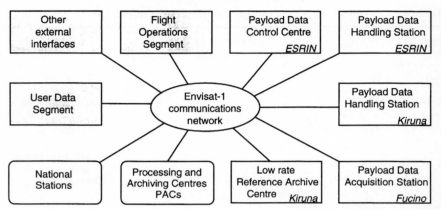

Figure 4.7 The Envisat Ground Segment Concept. Source: ESA (1994e)

4.5 EOSDIS

To support the EOS satellites, NASA is developing the Earth Observing System Data and Information System (EOSDIS). EOSDIS will provide command and control of the EOS spacecraft and will manage the data resulting from the EOS satellites, field measurement programmes and other data necessary for the processing and analysis of the EOS satellite data. EOSDIS will also process, store and distribute EOS data to US and other scientists around the world.

EOSDIS serves several roles within the NASA Mission to Planet Earth programme:

- NASA's Earth science data system for information management, archiving and distribution of NASA Earth science data;
- NASA's contribution to a broader network of data systems to support global change research;
- the main ground segment functions to support the EOS satellites.

EOSDIS therefore supports a large range of instruments on-board several satellites and a large range of worldwide scientists. As such it is unlike some Earth observation ground systems which are designed to control and process the data from a single mission. The architecture of EOSDIS has been developed to take account of changing science, technology, engineering and policy drivers, and in the process is liberating access to Earth observation data. Table 4.3 lists four trends that illustrate the development of easier access to Earth observation data.

A key element in this development is the recognition of the growing number of decentralised data sources and the growing number of decentralised, heterogeneous user bases.

Figure 4.8 shows an outline of the EOSDIS system. It is designed to handle data from both EOS satellites and satellites flown by international partners such as ESA and NASDA. The data ingest routes are both by direct reception and via data relay satellites. The data archiving system of EOSDIS is organised through eight Distributed Active Archive Centres (DAACs) which operate interactively to service user requests.

The EOSDIS core system will produce 260 standard products on a regular basis (Fox, Prasad and Szczur 1996) and will process and archive two terabytes of data per day. As well as Earth observation data, EOSDIS will also make available to its users a range of tools to process the data, such as document readers, source code compilers and visualisation tools, because the data system from the user perspective is only as good as the tools to access it. As preparation for the launch of the first EOS satellite, NASA has established a WWW site for users to access the early version (version 0) of EOSDIS. The site at

http://www-v0ims.gsfc.nasa.gov/v0ims/eosdis_home.html

provides access to the information management system for EOSDIS.

4.6 Centre for Earth Observation

While the ground segments for SPOT, ERS, Envisat and EOS are systems processing a defined and expanding set of Earth observation data, the European Commission Centre for Earth Observation (CEO) is a ground segment initiative to encourage a wider use of information generated by Earth observation satellites. The CEO recognises that users often have difficulties in obtaining Earth observation data, and yet the potential uses of Earth observation in the research, operational and commercial communities are large yet relatively underdeveloped so far.

The CEO concept has been developed in a different way from the main Earth observation ground segments because it has been developed by an

Table 4.3 Trends in EOSDIS System Architecture

Existing experience	EOSDIS themes
Product description and ordering	Product publishing and open access
Distinction of data and metadata	Seamless view of all data
Limited provider implementation	Extended provider implementation
Homogeneous, centrally managed system elements	Heterogeneous, autonomous system elements

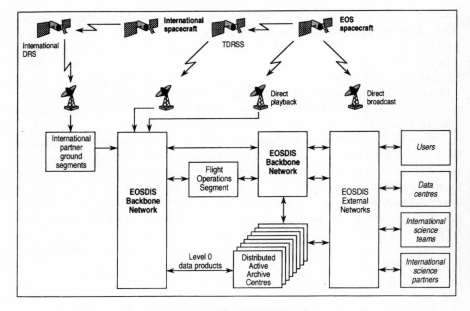

Figure 4.8 The EOSDIS System

organisation that is not responsible for any Earth observation missions, namely the European Commission, but which encourages wider use of Earth observation data. The strategy of the CEO is to provide applications support through projects and user support through case studies and user liaison, and to provide enabling services to help users to use Earth observation data.

The enabling services are comparable to some of the ground data processing systems of other organisations, but have been designed to exploit existing initiatives such as the Internet. One key component of the enabling services is the European Wide Service Exchange (EWSE) which allows users to connect to a wide variety of different Earth observation data and information sources. The main WWW site for the CEO at

http://www.ceo.org/

offers users a wide variety of information and services. Connection to the EWSE at this web site provides direct and indirect links to the following services:

- algorithms and software
- announcements and news
- catalogue searches

- data searches by location or keyword
- education and training in Earth observation
- product descriptions
- project and case study descriptions
- user descriptions

The EWSE is a testbed for the exchange of information about Earth observation amongst suppliers, users and data brokers.

With the growth of Internet access tools, such as search engines, this approach will build on existing ground segment investments and provide easier pathways to Earth observation data. The Earth observation sector is a good example of the Internet providing a substantially different capability to access data and services. In his book *The Road Ahead*, Bill Gates (1996) assesses the ways in which the Internet will enable easier and faster access to information for all. Earth observation will benefit from these Internet tools and services and will have the opportunity to become a core information service for all.

5 DATA DISTRIBUTION

5.1 Balance

Most of the investment in Earth observation goes into the space segment rather than the ground segment. For example, the Envisat-1 mission is costing about £1.5 billion, of which the ground segment cost is approximately £60 million. What the user sees of Earth observation is the data distribution side and this commonly receives one of the smallest proportions of the funding, although the US Earth Observing System is a notable exception.

Where data distribution is accomplished by direct broadcast, as is the case with Meteosat, GMS and GOES data, then the distribution is as efficient as the facilities which the users develop to collect and process their own data. However, for most Earth observation systems the users rely on the ground segments, such as those described earlier in this chapter.

SPOT is an interesting example because for the SPOT 1–3 satellites there is only one instrument on each satellite and a relatively limited list of products produced from that instrument. The ground segment is relatively centralised in that SPOT satellite control and programming plus much of the data downlinking is controlled from Toulouse, with the regional receiving stations interacting strongly with Toulouse. The SPOT catalogue system is efficient and effective at providing information on all the data held in the archive: new entries are included in the electronic catalogue the day after data acquisition by the SPOT satellite.

5.2 ERS

Geographical Restriction

The central control by the parent organisation, together with a good level of investment in data distribution support, has given SPOT an effective way of providing data to users and has given SPOT the largest sales volume of the high spatial resolution data providers. This is not the same story with the ERS programme.

The ERS data policy (ESA 1994c) contains a set of rules that govern the reproduction, distribution and sale of basic ERS data products by national and foreign stations. These rules are listed below. The ERS data policy document assumes that national and foreign stations will distribute only SAR data.

A. *ESA rights.* ESA shall retain the right to distribute any ERS data to any recipient worldwide but would, in the cases of C and D below, only exercise this right in the event of failure by the licensee to adequately perform their distribution duties.
B. *National, within ESA coverage.* For the distribution to residents of their own territory, operators of national stations shall be granted a non-exclusive license to distribute data collected by the national station within the coverage zone of ESA stations.
C. *National, outside ESA coverage.* For the distribution to residents of their own territory, operators of national stations shall be granted a non-exclusive license to distribute data collected by the national station outside the coverage of ESA stations.
D. *Foreign, outside ESA coverage.* For the distribution to residents of their own territory, operators of foreign stations shall be granted a non-exclusive license to distribute data collected by the foreign station outside the coverage of ESA stations.
E. *Other distribution.* Distribution, with the exception of the licenses granted under B/C/D above, will be under the authority of ESA, using distributors appointed by ESA.

These rules are clear in assigning distribution rights for national stations to *residents of their own territory*. In addition, ESA affirms the right to distribute ERS data to any recipient in any country, including those countries where this would place ESA through its commercial distributor in competition with the national station.

In 1991 the British National Space Centre (BNSC) and ESA signed an agreement to permit the reception of ERS SAR data at Britain's West Freugh national receiving station (ESA 1991). Accounting for the conditions for reception described above, ESA granted to BNSC, 'a non-exclusive license for

Physical access to Earth observation data   87

the reproduction, distribution and sale in the United Kingdom of those ERS-1 SAR data collected by the station'.

Combined with the main ERS data policy document, this restricts distribution of data received at West Freugh to residents of the UK. West Freugh can receive ERS SAR data from a coverage zone extending from Greenland in the north-west to Turkey in the south-east, but it can only distribute data to residents of the UK.

There is a freer condition for the distribution of analysed information, and the BNSC/ESA agreement notes that 'ESA grants BNSC a non-exclusive license, for the duration of the agreement, for the production, reproduction, distribution and sale of analysed information', without any geographical restriction. Analysed information is not tightly defined in the ERS data policy but appears to be the addition of external (non-ERS) information to produce new information products or services.

5.3 Standards

In both the CEOS and the US data policy principles there are statements concerning the use of international and national standards for data storage and data transmission. Much work has been carried out on standards in order to make access to data easier and to avoid reinvesting in producing methods of reading Earth observation data that already exist.

There are standards to meet most, if not all, situations for Earth observation data applications. One possible solution to the standards question is to define standards by government or quasi-government agencies and then for all users to apply these standards in their work. This is an attractive proposition because it means that all users can build on the existing experience of how to store and access Earth observation data tapes and CD-ROMs. This is useful for research scientists to make their work efficient because research groups can often work around technical problems, but it is of critical importance for commercial and operational users because if they are faced with a data tape or disc they cannot read then this is a substantial barrier which is difficult to overcome. For a research group a new tape format may be inconvenient; for operational and commercial users a new tape format may well put a halt to the data processing chain.

The results of the work on standards has, however, been a mixed blessing. As a UK system engineer once noted, 'the good thing about standards is that there are so many to choose from'. The development of standards may well reflect Gate's (1996) admonition: 'One area it's clear that government should stay out of is creating technology standards. . . Consumers, not regulators, should choose between compatibility and innovation.'

Earth observation is unlike the video tape market where one tape standard is dominant. Most suppliers have defined some form of standard for media and for formats, most of which are implemented differently by different organisations. Two different approaches to standards can be illustrated by the Committee on Earth Observation Satellites (CEOS) and the Consultative Committee on Space Data Systems (CCSDS). The CEOS approach is to agree methods of storing Earth observation data, for example by using explicit formats. This allows users of CEOS-format data to use CEOS reading routines to access the data. The CCSDS approach encourages suppliers to produce descriptions of the data formats on the data dissemination medium along with the original data. This means that data sets are self-describing and can adapt rapidly to changing technologies.

A key issue is the development of network and archiving technologies: these technologies are outpacing Earth observation initiatives in this area, and particularly the Internet and the tools available on the Internet are proving very attractive to the Earth observation community. Earth observation is a natural user of the Internet because (1) the data and processing tools are already in electronic form, (2) most users are computer literate, and (3) most users have been frustrated by the steps needed to find out about and then order Earth observation data.

Investment in the storage standards of (say) 1990 would be out of place for the technology of 1997, which in itself is likely to be out of place for the technology of the year 2005. A useful strategy for data storage and distribution is to stay in the middle of the most commonly used procedures, moving as the standards (if that is what they are) move.

This increasingly means using networks, which in turn places less reliance on physical media and greater reliance on electronic forms of data access. The growth of satellite telecommunications (e.g. the Hughes Spaceway proposed for 1998) will provide communications bandwidth on demand and so make Earth observation data dissemination easier and faster.

In the next 20 years Earth observation will be best served by relying on generic data storage and networking tools available either in the commercial domain or as freeware on the Internet. These technologies will enable and encourage different modes of access to Earth observation data, which will assist in the maturity of the sector.

5.4 Trend

Growth in demand, increases in the number of applications and easier electronic dissemination over the Internet provide a changed context for Earth observation data distribution. What is needed now is greater flexibility in data access, which means an increased consideration of direct broadcast and a lower level of control of data distribution. There is a need to control or police the right to have access to Earth observation data, but there is a countering need to liberalise the distribution of data to users who have the

Physical access to Earth observation data   89

right of access. Liberalisation of distribution and more innovative uses of networking technologies will assist the growth in applications by allowing more innovative intermediate services to be developed.

6 ANNOUNCEMENTS OF OPPORTUNITY

6.1 AO Instruments

Many Earth observation systems provide opportunities for flights of instruments provided by organisations other than the satellite operator. These instruments are typically termed Announcement of Opportunity (AO) instruments. As shown in Table 4.1, on Envisat-1, for example, the instruments are divided into ESA-developed instruments and Announcement of Opportunity instruments provided by organisations in the ESA member states.

Schemes for AO instruments are beneficial for both the AO instrument provider and the satellite operator. For the instrument provider it provides a means of flying an instrument without the need to build a dedicated space mission. For the satellite operator it provides a means of enlarging the data collected by a mission without incurring the investment costs in developing an instrument.

AO instruments do, however, lead to tensions in the Earth observation community which arise from data policy problems. A key question is: who owns the data? In the case of Envisat-1, ESA owns all the data from the spacecraft including the data from the AO instrument. This may seem invidious to the instrument provider because it is the organisation own instrument built at its own cost to provide the organisation with the data it requires for its own purposes.

Even if AO instrument providers accept that ownership of the AO instrument data rests with the satellite operator, there is often a claim that the data should only be accessible to the instrument provider, or accessible first to the instrument provider.

The tension comes from the desire to have a broad spread of all Earth observation data and to lessen rather than widen the restrictions of the use of the data. Scientists involved with AO instruments have sometimes claimed that it is only they who possess the understanding of the calibration of the instrument and the model applications in which the data can be used, whereas the wider scientific community (ESF 1992) believes that access to data should not be fettered by institutional control, but that scientific progress is achieved by use of the Earth observation data and open, peer-reviewed publication of the results.

6.2 AO Projects

Announcement of Opportunity projects are initiated by satellite operators to stimulate innovative applications of Earth observation data. Data are typically provided to the AO users free of charge. ESA, for example, has had several calls for AO projects for the use of ERS data, and has awarded about 600 of these AO projects.

There is a delicate balance to be struck on AO projects. On the one hand, AO projects develop new and innovative applications of the data, particularly when the data are of a new type such as SAR data. On the other hand, the data are provided free of charge and this results in two problems: first, the provision of free data inhibits what might otherwise be sales of the data; and second, the focus is on the technology of data exploitation rather than the end-user of the data. The challenge is to achieve the right balance between stimulating applications that illustrate the benefits of the data and avoiding the considerations of data sales that require end-users to build into their investment plans the costs of the Earth observation data.

5
Data Protection

1 INTRODUCTION

The large investment in acquiring Earth observation data has led to a concern for protecting that investment by providing some form of protection for the data (Dreier 1992). The legal instruments used in Earth observation data protection are neither simple nor uniform, and there is no single, clear-cut legal protection that covers all Earth observation data (Thiebault 1992).

The purpose of this chapter is not to analyse the legal issues in detail, but to review those legal issues which have data policy implications. The early part of the chapter discusses the general framework and the later part gives a specific example of encryption by EUMETSAT to protect its data from unauthorised use.

2 LEGAL PROTECTION OF DATA

2.1 The Need for Legal Protection

The issue of the legal protection of raw Earth observation data is concerned with the establishment of property rights or rights of ownership over data sets by the organisations that generate the data. For example, the data policies of both ESA and EUMETSAT refer to such rights. ESA sums up the position of many satellite Earth observation supplier organisations when it notes (ESA 1996a), 'For the Agency it is important to have a clear legal basis for protection of the data in order to strengthen the regime for the ERS-1 data reception Agreements and the Agency position in the negotiations in international fora with regard to the exchange of data and data policy in general.'

Two main questions arise in connection with this issue:

- Is the legal protection of data necessary and justified?
- If so, what are the possible mechanisms for protecting the data?

It could be argued that Earth observation data should be made available as widely as possible, free of charge. Under these circumstances there would be no need for legal protection of the data since such protection would serve no purpose.

The protection of data is necessary, however, in the case of suppliers (whether public sector or private sector) requiring a return on their investments in the systems generating the data. Protection is required in order to allow the distribution of data to be controlled, so that either revenues can be achieved from their sale or resale, or when the data are made available at no charge to the user (as is the case with principal investigator programmes) their use and onward distribution can be controlled. Without adequate protection, data could be passed from one organisation to another, particularly given easy electronic access, without the original supplier receiving any revenue or having any control over the distribution of the data.

In this sense the issues for Earth observation are comparable to the broader issues in intellectual property protection. Large initial investments are characteristic of the production of much intellectual property (e.g. films or pharmaceuticals), and downstream reproduction of the material is low cost. Legal protection of intellectual property, including Earth observation data, is closely linked to issues relating to pricing policy: these pricing policy issues are discussed in Chapter 6.

A simple legal mechanism for protecting Earth observation data from unauthorised distribution is the inclusion of appropriate restrictions in the contracts on the basis of which data are sold. The problem with this, however, is that if the contract is breached there is no protection against further dissemination of the data. Consequently, much attention has been paid to the quest for a more direct legal basis for protecting data. The main solutions that have been proposed (Gaudrat 1992a, b) are discussed below.

2.2 Copyright

One of the main mechanisms for the protection of Earth observation data is copyright. There is no international law on copyright and so such protection must be granted by national copyright laws. These laws cover many sectors, including geographical information (EUROGI 1996) as well as strictly Earth observation data.

European national laws are mixed on the subject of copyright (Gaudrat 1992a), while the US position is clear (Shaffer 1992): '[The US] Government is prohibited from copyrighting data generated at public expense. Data from privately-funded systems may be protected through copyright.'

There are two main technical issues which are debated in copyright protection for Earth observation data, namely fixation and creativity. Some national laws give copyright protection when data are fixed in some form. It has been argued that fixation occurs when a satellite sensor measures the electromag-

Data protection

netic radiation that passes through its aperture, however briefly. In the case of CCD arrays there is definitely a short-term fixation because this is how the sensor works. In the case of other sensors, such as scanning radiometers, a period of fixation is harder to justify.

Even if there is difficulty in establishing fixation at the sensor, Earth observation data are fixed in the sense that they are recorded onto a magnetic medium either on-board the satellite using on-board tape recorders, or at the ground receiving station where there is typically some form of magnetic storage.

The issue of creativity is more difficult to resolve. UK copyright law requires a certain degree of skill and labour to be eligible for copyright, whereas French law is stricter and requires an element of originality to qualify for creativity (Thiem 1992). Dufresne (1992) argues that SPOT data are covered by French copyright law because their creation fulfils the criterion of originality in French law: 'one can say that a work is original when it has been created through a process which implies personal and subjective choices from its author'.

The generation of SPOT data requires complex programming by the ground teams at the satellite control centre in Toulouse, including the red and blue programming discussed in Chapter 4. The control centre teams are the true authors of the SPOT views. Dufresne (1992) argues that such complex programming implies that the programmers make subjective choices in selecting how to point the SPOT instrument and when to turn it on or off. These subjective choices mean that the SPOT ground teams are creating original works in producing SPOT images.

There are two extra considerations in the use of copyright. The first concerns the definition of what constitutes Earth observation data. If Earth observation data are merely information, then as Dreier (1992) argues, 'Under copyright, however, information is said to be free and cannot be monopolized. Traditionally, it is only the (visual) representation of information which may qualify for copyright protection.' The second consideration is that copyright does not protect against the reception of Earth observation data but the onward reproduction and use of the data.

Copyright is an important issue in the long term to provide legal protection of Earth observation data and it is very valuable to create clear guidelines for both users and suppliers on what constitutes copyright protection. However, there is a danger that some national laws may allow users to avoid copyright issues altogether. Under the US Freedom of Information Act (FOIA), it is claimed that: 'data in the physical possession of US Government employees or at US Government facilities must be made available to any requestor under FOIA at the cost of reproduction regardless of whether the data are also available through a commercial distributor' (Shaffer 1992). This condition could potentially create problems over the distribution of SPOT data and

Meteosat data, which are explicitly protected by CNES and EUMETSAT respectively.

2.3 European Directive on Databases

Objective

In March 1996 the European Parliament and Council agreed a Directive on the legal protection of databases (European Commission 1996a). The objective of the Directive is to 'afford an appropriate and uniform level of protection of databases as a means to secure the remuneration of the maker of the database'.

The Directive covers any relevant database (e.g. census data or topographic data), and its scope is defined as follows: 'For the purposes of this Directive "database" shall mean a collection of independent works, data or other materials arranged in a systematic or methodical way and individually accessible by electronic or other means.' On this basis the Directive does appear to cover Earth observation data, even noting the later exclusions in the Directive.

In the preambles there are two items out of the 60 issues of justification which appear to be directly relevant to Earth observation data. The first notes that in Europe no clear legislation exists to protect databases: 'Whereas databases are at present not sufficiently protected in all (European) Member States by existing legislation.' And the second relates directly to one of the major international tensions in Earth observation data policy: 'Whereas the making of databases requires the investment of considerable human, technical and financial resources while such databases can be copied or accessed at a fraction of the cost needed to design them independently.'

The Directive proposes protection of databases by either copyright protection or by a *sui generis* protection.

Copyright

The core of the copyright protection in the Directive is the selection or arrangement of the database rather than protection of the contents of the database. Under the Directive, copyright is used to protect an author's own intellectual creation of the organisation of the database. Therefore the issue of copyright protection under the Directive revolves around the issue of who is responsible for the organisation of the database.

As noted earlier in this chapter, the issue of intellectual creation is not without questions in Earth observation. The author in the context of the Directive is an individual or group of individuals, so it is always possible to

Data protection

associate the teams of people who create the software to control a satellite and its instruments with the term 'author' in the Directive. A good case in point is SPOT, but in other Earth observation systems the level of the author's control of the organisation of the database may not be so easy to identify.

Sui generis *Right*

The Directive also provides for a *sui generis* right, i.e. a right 'for the maker of a database which shows that there has been . . . a substantial investment in either the obtaining, verification or presentation . . . of the contents of that database'.

This part of the Directive concentrates more on the contents of a database rather than the author, and so might be used to protect Earth observation data themselves. Interestingly the *sui generis* right allows a lawful user the right to extract insubstantial parts of the contents of a database and use the extract for any purpose whatsoever. This opens the door to organisations to extract small parts of Earth observation data sets, depending on interpretation of the term 'insubstantial', and use them in any way they wish.

Exceptions

For both the copyright and the *sui generis* right there are exceptions. One exception which relates to Earth observation data policy is for scientific research: the Directive notes that Member States may agree legislation to allow special conditions for scientific research that uses electronic databases as long as the source is indicated and there is a definition of the scope of non-commercial use.

A second exception appears to be in contradiction with the whole idea of protection: 'The first sale in the Community (i.e. European Member states) of a copy of the database by the rightholder or with his consent shall exhaust the right to control resale of that copy within the Community.' For Earth observation this is interesting because it appears that there could be a loss of control of the distribution of copies of Earth observation data sets. This part of the Directive could be used to support arguments both for greater control and for less control of Earth observation data.

In the Directive the term of protection foreseen for databases is 15 years, either from the creation of the database or from the date the database is made available to the public. For Earth observation this typically means 15 years from the date of data capture since the potential use of the Directive for Earth observation data is based upon the premise that an identifiable, electronic database has been created either on-board the spacecraft or at the ground station which receives the data.

2.4 Legality of Price Discrimination

One aspect of Earth observation data pricing policy is the possibility of applying different prices to different users for the same product. In particular, there is a widely held view that research users should be able to obtain data at preferential prices provided they follow certain conditions. In this context it is important to consider the legal issues related to price discrimination.

In Europe and elsewhere, price discrimination can be prohibited either on the grounds of anti-competitive behaviour or on the grounds of abuse of economic power. Almost all European national competition laws contain such provisions, as does the EC Treaty of Rome (Articles 85 and 86).

Price discrimination can be prohibited on the grounds of anti-competitive behaviour if the discrimination is applied between entities that are in competition with each other. This means, therefore, that Earth observation data could be legitimately supplied to research institutes for research use at lower prices than to commercial enterprises since the two types of entity do not normally compete with each other. This position only holds provided the research institute does not use the data for any commercial activity (e.g. creation and sale of derived products or the use of the data in consultancy projects), thereby entering into competition with commercial companies.

Price discrimination can also be prohibited on the grounds of abuse of economic power. This can apply either in the case of an individual enterprise with a dominant position in the market, or in the case of a group of undertakings which jointly hold a dominant position. Because investment costs are so high, there are relatively few Earth observation data suppliers in the world, which means that it would only require an agreement on prices between a small number of suppliers to create a situation of market dominance.

The prohibition on price discrimination cannot apply, however, if it relies on objective differences between the entities subject to discrimination and does not create any disadvantage in the competitive market. For instance, if data are provided to research users at preferential prices in return for publication of results in the open literature, then this would be regarded as an objective criterion and so there would be no prohibition of price discrimination.

Earth observation data can therefore be offered to certain categories of user or for certain types of use at preferential prices, provided the supply of the data is made subject to certain conditions aimed at precluding unfair competition.

3 ENCRYPTION

3.1 Data Reception

One way to protect Earth observation data is to encrypt them so that only those receiving stations with the decryption key can gain access to the data. Data from EUMETSAT's Meteosat satellites can be received at either Primary Data User Stations (PDUS) or Secondary Data User Stations (SDUS). The PDUSs receive data in digital form, whereas the SDUSs are only capable of receiving analogue transmissions of poorer spatial resolution and image quality.

Until 1995 both PDUS and SDUS receiving stations were able to receive Meteosat data in an unencrypted form. From 1 September 1995 most of the digital transmissions of Meteosat data were encrypted and could only be unencrypted by those PDUS users who held the decryption key. The decryption key is provided by EUMETSAT to registered PDUS users (about 200 in 1996), who follow certain procedures. This section describes the characteristics of the process of protecting Meteosat data by encryption.

3.2 Rationale

The rationale behind EUMETSAT's decision to encrypt some of its data transmissions is based on the following three points (EUMETSAT 1994a,b):

- EUMETSAT is a voluntary cooperation of the National Meteorological Services (NMSs) of 17 European states. These states collectively contribute approximately 200 million ECU per annum to fund EUMETSAT. If any country not a member of EUMETSAT can receive all data free of charge, then what is the motivation for that country to belong to EUMETSAT and to pay its contribution to the EUMETSAT costs? Control of the data by encryption provides the mechanism to control access to the Meteosat system.
- There is a growing trend towards commercial meteorological services in both the public and the private sectors. If the commercial services do not contribute to the funding of the Meteosat system then why should they be granted access to the data at no charge? Again, control of the data by encryption provides a mechanism to control access.
- A better identification of the users of the Meteosat system will allow EUMETSAT to become more informed about its user base and better able to respond to user needs. The introduction of encryption has been accompanied by an improved registration procedure which increases the information about the user base.

3.3 Access to Data

Based on this rationale, EUMETSAT introduced on 1 September 1995 encryption of some of the Meteosat High Resolution Image (HRI) data products for reception at the Primary Data User Stations. The poorer quality data products transmitted to the Secondary Data User Stations were unaffected.

The six-hourly HRI data remain unencrypted and are available at no charge to all users who possess a Primary Data User Station. The six-hourly transmissions are at 0000, 0600, 1200 and 1800 hours GMT. The three-hourly HRI data are encrypted, but free of charge to all NMSs that can receive the data.

The hourly and half-hourly HRI data are encrypted. The price for access to these data depends upon the relative wealth of the countries outside the EUMETSAT member states. For the NMSs of those countries which have an annual per capita Gross National Product (GNP) of less than US$2000, the data are free of charge for use within the NMS. This also applies to the NMSs of those countries subject to tropical cyclones. NMSs of wealthier countries (above US$2000 annual per capita GNP) and other users have to pay an access fee.

In order to decrypt the encrypted HRI data, a PDUS has to include a decryption unit integrated into the PDUS, to which a separate Meteosat Key Unit is connected. The Meteosat Key Unit is provide by EUMETSAT to authorised users who register with EUMETSAT. Thereby EUMETSAT has direct control over which encrypted data can be collected by a PDUS and a register of all the PDUSs collecting data with a frequency of greater than six hours, because without a Meteosat Key Unit a PDUS cannot receive the three-hourly, hourly and half-hourly Meteosat data.

4 TRENDS

The trend in Earth observation data protection is towards clearer definitions of the control of access. There are firmer foundations for the legal protection of data and access to data through the European Commission Directive on databases and through the US Freedom of Information Act, although the intentions of the two are different.

EUMETSAT provides one example of physical data protection by data encryption. Because it is relatively easy to copy electronic information, there is a growing interest in physical techniques to prevent copying of tapes and CD-ROMs that hold Earth observation data. It is technically possible to include date information on digital databases that restrict access to certain time periods: once the period is completed then access to the data is denied. This is comparable to evaluation licences for software.

Data protection

Even when a CD-ROM cannot be physically protected, it is possible to insert a digital watermark into the data so that any data or subsequent information copied from the database will continue to carry the watermark or fingerprint showing its origin. This will make enforcement of legal protection easier to achieve.

The main trend is one of attitude. Many in the Earth observation sector, particularly in the research community, believe that Earth observation data should be free of charge. After all, you cannot see the data when they are held on a disc, and the data are of our own environment which we all feel we can readily access. This approach leads to leakage of data and copying of data in electronic form against the limitations of licences. One can draw an analogy with computer software. Software cannot be seen in its raw form and it can be easily copied. In the early days of software on personal computers there was extensive software piracy by taking electronic copies. Because of policing by the software industry there is now a clear recognition that software running on all computers must be properly licenced, and a far smaller percentage of software users employ unlicenced software. This change was achieved by prosecutions and by raising the profile of the issue. Earth observation is in the equivalent of the early stage of personal computer software. The sector must move to the mature phase where the attitude of Earth observation data users is to respect the licence conditions of the data they are using. This may mean that data are free of charge, as is the case with EOSDIS products, or that data have a price and sometimes other access conditions. Explicit agreement with all the licence conditions will become the norm for Earth observation, as it has with software, to the overall benefit of the sector.

6
Data Pricing Policy

1 INTRODUCTION

Much of the international debate about Earth observation data policy eventually returns to the theme of pricing policy. Whether it is legal protection of the data, distribution rights or return on investment, pricing policy often takes centre stage in the Earth observation data policy debate (Macaulay and Toman 1992). The purpose of this chapter is therefore to examine the issues and options concerning pricing policy for Earth observation data.

Earth observation data pricing policy is subject to forces or pressures from many directions. There is widespread agreement on exploiting to the full the potential offered by Earth observation (Shaffer and Backlund 1990), but the views on pricing policy vary depending upon the balance of the main forces on pricing policy. Figure 6.1 illustrates these forces in summary form (Harris and Krawec 1993b).

Four forces are identified in Figure 6.1: the value of data, the role and influence of suppliers, the role of users as buyers, and substitutes for Earth observation data. The value of Earth observation data is fundamental in determining an appropriate pricing policy, although the approaches taken to value differ in the research, operational and commercial areas. The relationships between suppliers and users are complex and interconnected because much of the funding for Earth observation systems and applications originates in government. The growing emphasis on the role of the user in Earth observation means that more resources need to be made available to the user community. A rebalancing of government funding in favour of research and operational users would give greater power to users and so assist in guiding the development of the sector.

Earth observation data are in a competitive market in the broadest sense. There are substitutes for satellite Earth observation data, both within the sector and outside the sector, for example:

- agricultural field radiance data collection by using radiometers on-board helicopters

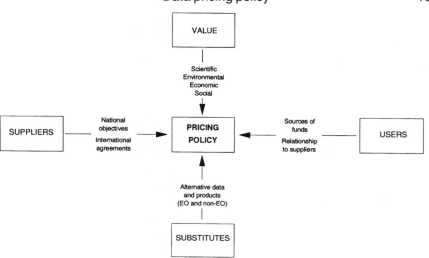

Figure 6.1 Forces Acting on Earth Observation Data Pricing Policy. Source: Harris and Krawec (1993b)

- agricultural monitoring by field visit and inspection
- digital terrain model generation by stereo aerial photography
- oil slick identification from the bridges of ships
- topographic mapping by ground survey
- urban growth mapping from population statistics

Significant expansion of Earth observation will occur when new uses of the data are identified, and this expansion is likely to occur when Earth observation data substitute for existing data sources as well as providing new data. Users of Earth observation data will switch from one data source to another if the pricing policy is perceived to be a barrier to data exploitation.

The forces shown in Figure 6.1 evolve through time. In the short and medium term a key influence on Earth observation pricing policy is the debate on whether the supplier pays or the user pays. In the longer term the widening of the definition of the value of the data will alter the balance of the pricing policy debate.

2 POLICY FOUNDATIONS

Before analysing the pricing policies themselves, the early part of this chapter discusses three main policy foundation issues: first, the fundamental requirements underlying Earth observation data pricing policy; secondly, the basis for the pricing of Earth observation data; and thirdly, the preferential treatment offered to certain categories of users.

2.1 Fundamental Requirements

Introduction

The right pricing policy in the right circumstances will have a significant impact on the successful exploitation of Earth observation data in the research, operational and commercial sectors. But why is pricing policy such a problem and the subject of such long-running debate? There are three fundamental requirements that need to be taken into account when addressing pricing policy, and these are the capability of organisations to reach their declared objectives, the return on investment in Earth observation systems and the need to ensure access to data for certain key categories of use.

Reaching Objectives

There is a clear need to match Earth observation data pricing policy with the objectives of the missions, systems and organisations that participate actively in the sector. One fundamental objective of many Earth observation organisations is to maximise the public benefit of the large public investment in Earth observation (European Commission 1992a). However, this objective cannot be specified in detail using solely quantitative criteria, and so must be characterised by qualitative criteria such as improved understanding of global change, better climate models or improved environmental forecasting systems. In economics this is termed the improvement of economic welfare. The French interministerial policy discussed in Chapter 3 is an example of this.

Because Earth observation is partly commercial (e.g. Space Imaging and SPOT Image), there is also a need, when assessing the public benefits realised from public investment, for an effective blend of government actions and market forces. The UK civil space policy is a case in point (BNSC 1996).

Earth observation data are sources of information about phenomena that are public in their nature, e.g. climate change and natural resources. The data are also sources of information that have private value, e.g. coastal change, timber logging, commercial agriculture and ice forecasting. Because of this mixed character, Earth observation data are neither exclusively the responsibility of the public sector nor of the private sector. One consequence of this duality is that each sector recommends that more financial resources be provided for Earth observation by the other sector. An Ordnance Survey (1996) report sums up this problem in the broader context of geospatial information: 'when there are "external" benefits from the provision of information and both public and private sector users with similar but not identical needs, how should the "fixed" costs of [data or information] provision be shared between them?'

Meeting objectives within a single national or multinational organisation may not be possible by using only the data from that organisation. For example, each National Meteorological Service acquires and uses meteorological data (via the Global Telecommunications System) that originate in other meteorological services. There will be a need to exchange Earth observation data from different sources and this process will benefit from compatible data pricing policies.

Return on Investment

The finance departments of most governments are increasingly trying to ensure that the user pays rather than the taxpayer pays wherever only subsets of the population benefit from government investment, and many governments are now recognising the need to recover part of their financial investment in Earth observation systems. The mechanism to achieve cost recovery in Earth observation is a data policy that recognises this legitimate requirement but is sensitive to the desire also to maximise return on investment in a non-monetary sense. For example, scientific research and operational monitoring that use Earth observation data return value to governments. This value may not be easily expressed in financial terms, but is clearly a return on Earth observation investment.

There is a distinction that can be drawn between the nature of the returns on investment. For operational, scientific, social and political areas the maximum return on investment can be partly met by maximising use of the data. This means that the likelihood of maximising value is increased. For the commercial area, governments wish to see recognition of the investment by them in Earth observation systems, but also acknowledge that the financial return on investment in the short and medium term cannot recover the investment in the space and ground segments, and is more likely to produce a financial return on only the running costs of Earth observation systems. This distinction in the nature of the argument concerning return on investment is a result of the various government attitudes to one central question: who pays for data creation and distribution? Governments have approached this question differently.

Key Categories of Use

One fundamental requirement of many nations and organisations involved in the sector is to maximise the research use of Earth observation data. For example, the purpose of the US policy statements on data management for global change research is 'to facilitate full and open access to quality data for global change research' (NSTC 1996).

Many research programmes in global change do and will require access to large volumes of Earth observation data, but the research community typically has only a limited ability to pay for Earth observation data. It has been estimated that in the USA a typical global change research grant is approximately $150 000 for one year, of which about $20 000 may be available for the purchase of data (S. I. Rasool pers. comm.). This would allow, for example, the purchase of four Landsat Thematic Mapper geocoded digital scenes, which is unlikely to be adequate for a global change research project.

Because the achievement of high quality science is a major objective of much public investment in Earth observation, the successful resolution of the question of Earth observation pricing policy for research use is a fundamental requirement of many funding agencies. The same may be said of the operational use of Earth observation data for the public benefit on a non-commercial basis. Issues concerning access to data for these two key categories of use are discussed in a later section.

2.2 The Basis for the Pricing of Earth Observation Data

Having discussed the fundamental requirements behind Earth observation data pricing policy, this section discusses the basis for the pricing of Earth observation data.

Economic Characteristics

Earth observation data provide information on the environment which have a potentially wide use. The data are therefore comparable to general-purpose geospatial data such as environmental statistics, topographic surveys and land registration data. Geospatial data have the following economic characteristics (Ordnance Survey 1996).

- high, fixed collection costs and low, variable reproduction and dissemination costs
- non-rivalry in consumption
- many different uses and markets, and the same information may have different value in different markets
- additional 'external' benefits such as improvements to policy making and discoveries of substantial environmental change
- new technology, particularly access to the Internet, is changing the underlying business economics of the user community

Many governments are recognising the necessity of recovering part of the costs of Earth observation missions. In closed loop systems the users collectively fund all the costs. In open loop systems only part of the costs are recovered from the users. Typically, the costs recovered are those attributable to parts of data processing, archiving, dissemination and customer service, leaving the investment in the space segment, the launch and ground control not recovered. If only a part of the costs is recovered then this raises the question of how to set pricing levels.

The basis for cost recovery can be assisted by economic theory. A number of pricing conditions and approaches (Ordnance Survey 1996) can be envisaged for all information:

- A competitive market in which there are a large number of well-informed buyers and sellers. A stock market fits this description, but not yet Earth observation.
- Dissemination cost pricing where the price is set at the marginal cost of dissemination. Where dissemination cost pricing is supported by government subsidy, as is the case for Earth observation, this results in a suboptimal allocation of resources since the government subsidy is normally greater than the benefit to users of the lower price achieved by dissemination cost pricing.
- Classic monopoly with only one supplier, which is not the case with Earth observation.
- Natural monopoly where the very high fixed costs and low marginal costs make it inefficient for more than one firm to supply the market because of the duplication in fixed costs involved. This situation is interestingly characteristic of Earth observation but there is not a natural monopoly; rather there are many suppliers (SPOT, Radarsat, ESA, NASA, NASDA, etc.) all with very high fixed costs, low marginal costs and small markets.
- Price differentiating monopoly where the fixed costs of supply are recovered from higher value users, but users with a lower willingness (ability) to pay are charged a lower price. This approach could be characteristic of Earth observation, although there are substantial problems in defining value to the user.

Maximum Use

One of the key concerns of many organisations involved in Earth observation is to maximise the use of Earth observation data. For example, the European Commission (1992b) has recognised that 'the main underlying goal which an Earth observation data policy for Europe would aim to achieve . . . is to maximise the public benefits realised through the large public investments which are made in satellite Earth observation systems.'

Shaffer and Backlund (1990), writing from NASA, propose that the 'goal of

US Earth science data policy should be to provide the widest possible dissemination of the data'. Maximising use of Earth observation data has also been explicitly recognised by the European Science Foundation (ESF 1992) and by CEOS (1994).

If the objective of public investment in Earth observation is to maximise the public benefit from exploitation of the data then the widest possible use of the data is one mechanism to achieve this objective. Using maximum use as the basis for the pricing policy for Earth observation requires the securing of low barriers to the exploitation of the data. Low barriers do not mean that data are provided at no cost: the experience of the Centro para el Desarrollo Tecnologico Industrial in Spain indicates that even when data are offered at no cost to users the take-up is very limited. Where barriers do need to be low is in the arrangements for exchange and sharing of data so that the value of the information is spread as widely as possible.

Value of the Data and Closed/Open Loop Systems

The reason for society to engage in Earth observation is to bring value to society from its investment. But what is meant by value? In the research area the value of the investment in Earth observation can be stated as providing to the community at large, facts or principles concerning the planet Earth. This may take the form of discoveries (e.g. the rate of forest change in tropical regions), or it may take the form of demonstrating or validating the geophysical meaning of space measurements.

In the operational area the value of Earth observation lies in improvements to the capability to monitor and predict the behaviour of the environment, e.g. weather forecasting or hazard warnings. In the commercial area the value of Earth observation lies in enabling or improving services provided to end-users (e.g. crop yield prediction and marine oil exploration), which means that benefits such as reducing costs or improving service must be explicitly expressed. Value can therefore be expected to have an impact on the basis of the pricing of Earth observation data, and this in turn has implications for funding structures.

The debate on pricing is often expressed as whether the user should pay or the supplier should pay. In closed loop systems the suppliers and the users are the same, but in open loop systems the suppliers and the users are different. Therefore the nature of the system (closed or open loop) will have significance for the debate on value and pricing policy. For closed loop systems the value of the data is defined before the investment is agreed. In open loop systems the investment decision is made on the basis of anticipated value. Examples of closed and open loop structures are presented in Figure 6.2.

Figure 6.2(a) illustrates the closed loop system operated by EUMETSAT.

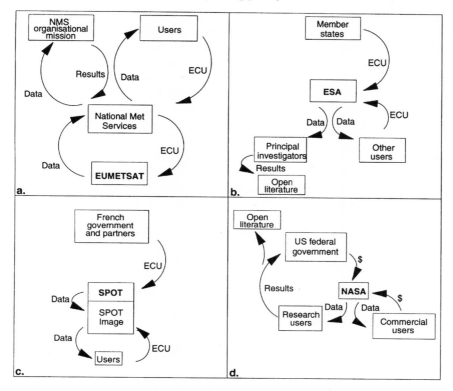

Figure 6.2 Examples of Open and Closed Loop Systems

EUMETSAT is funded by its European National Meteorological Service partners which receive and then use the data for their own internal purposes, and can also distribute data and products to other users within their own territories at a charge to those users.

Figure 6.2(b) illustrates the open loop system operated by the European Space Agency. The ESA member states provide funds for ESA which then provides data to principal investigators and pilot projects, and also sells data to other users. The principal investigator or other user may or may not be linked to the organisation within a member state that provides the initial funding.

A national open loop system is illustrated by France and the SPOT system (Figure 6.2(c)), where SPOT Image exploits the investment by the French government and its Belgian and Swedish partners in the SPOT satellites by selling SPOT data and products to all users at the same price and on a commercial basis.

Figure 6.2(d) illustrates a closed loop system with an open loop extension. Because NASA uses federal government funds to support both data capture (space and ground segments) and data exploitation (research use), this is a closed loop system. It has an open loop element in addition because data are also sold separately on a commercial basis.

In a closed loop system, such as that of EUMETSAT, the primary user pays for the whole data capture system. There is no charge for data at the point of delivery to the user because the primary user has already paid for the system. In a closed loop system the value of the investment is therefore assessed by the users who continue to pay for the system based upon the benefits of the investment. One advantage of this structure is that because users pay all the costs for the data they have a strong influence over the design of the programme and the Earth observation system will better meet their needs.

In the case of an open loop system, such as that of ESA, the situation is characterised by the supplier paying a relatively large proportion of the cost of providing data because the price to the user of data and products is relatively low. In an open loop system the perceived value of Earth observation may depend more on what is funded from which source than on an objective assessment of final results.

2.3 Access to Data for Key Categories of Use

Relevant Categories of Use

In general, preferential treatment is offered to the research community and those operational organisations that operate for the public benefit on a non-commercial basis. The latter category essentially encompasses meteorological services at present, but with a growing international interest in environmental monitoring the category may grow significantly to include organisations responsible for monitoring river pollution, climate and land surface hydrology.

Research users receive preferential treatment because, as noted above, it is recognised that their willingness or ability to pay is limited compared to the charges made for data. Research users are commonly funded by governments which themselves fund Earth observation systems, so a preferential rate improves the likelihood of maximising the use of Earth observation data. Research users also return value to society at large by discoveries and innovations that are published in the open literature.

Non-commercial users acting for the public benefit receive preferential treatment because of their public service. Accurate forecasting of extreme weather hazards by meteorological services, for example, provides society with information that has direct economic and social benefits.

Access Conditions

The policy statements that provide for preferential treatment for research and operational categories have public good benefits at their core. Given that there are pricing conditions which favour certain categories, how are these categories defined and are there problems with the definitions?

Definitions of research use. The suite of US policy statements on data management for global change research recognises that research users are defined differently by different US agencies, and because there are 18 agencies in the US Committee on Earth and Environment Sciences (CEES) which endorsed the policy statements, there is scope for wide interpretation. The IEOS principles suggest that agencies designate research users through an Announcement of Opportunity or similar formal review mechanism. Definitions such as these can be interpreted to encompass research use in both the public and the private sector, and the boundaries of the definitions are not clearly defined.

In both the USA and Europe there is concern that research users will obtain data at lower prices for research use and then use the data for commercial consultancy purposes, i.e. the leakage problem. This concern is not diminished by the generally loose definition of research, and such actions would constrain the development of a mature commercial sector.

There are also legitimate research users not covered by some of the definitions of research, so they cannot claim membership of the research category for data at the lowest possible cost. Those carrying out blue sky research funded, for example, by a university or not-for-profit research institute may not be able to qualify for low cost data because they do not meet the specification of an Announcement of Opportunity or similar process. This is iniquitous and does not support the objective of maximising the use of Earth observation data.

Research use is often carried out in educational institutions, but there is no category of preferential treatment for education and training use of Earth observation data. Research use is research use and not a broader educational category.

Description of publication. One part of the description of research use is the publication of results, data and algorithms. For example, the fourth IEOS principle states that 'Research users shall be required to submit their results for publication in the scientific literature.' The underlying principle here is to provide value to society by open publication and access to all of the results of the scientific research. However, what constitutes submission of results for publication in the scientific literature? *Nature* and the *International Journal of Remote Sensing* will qualify, but will submission to *The Economist* be outside the scope?

If results of scientific research are limited to tables and graphs then it is highly unlikely that scientific journals will accept these by themselves. Monographs are an alternative, but their popularity is in steep decline and they are not a serious proposition. If the criterion for the research category is that researchers will submit their results for publication in the scientific literature, then this is a very unclear definition of the category.

Marginal cost price. The marginal cost price is a term that is widely used to describe the lowest possible price for certain categories of user which receive preferential treatment. This is the same as the dissemination cost pricing used by economists and described earlier. The heart of the marginal cost argument is that users in the preferential categories should not be required to fund, through charges for data, the costs of the core Earth observation space and ground segments.

The terms used for marginal cost have included the marginal cost of reproduction and delivery and the marginal cost of filling a specific user request. However, are these adequate descriptions of the intended price level? What would happen if a major research project or operational system (e.g. an IGBP project or GCOS activity) required a sufficiently large increase in Earth observation data that a supplier organisation had to increase its data handling computer capacity by 25%? Would the cost of filling a specific user request encompass the cost of providing the extra hardware and software?

This also introduces the concepts of short-run marginal cost and long-run marginal cost. The short-run marginal cost is taken as the extra cost to provide data beyond the investment in the basic ground infrastructure, while the long-run marginal cost includes the funding necessary to modify and develop the ground infrastructure in response to changing needs. Different agencies interpret marginal cost in different ways depending on their view of short-run or long-run marginal cost.

3 PRICING POLICY OPTIONS

3.1 Structure

The first part of this chapter has discussed the policy foundations for Earth observation data pricing policy. The discussion now turns to those pricing policy options (Harris and Krawec 1993b) which have been used and proposed in practice, and which, either implicitly or explicitly, incorporate the issues that have been discussed in the first part of the chapter.

The seven data pricing policy options listed below are discussed in this part of the chapter:

Data pricing policy

- free data for all users;
- marginal cost price for all users;
- market-driven, realisable prices for all users;
- full cost pricing;
- two-tier pricing;
- information content pricing; and
- rebalancing of government funding.

Each of these options is discussed using the same four-part structure: a definition of the term; arguments that have been proposed in favour of the option; arguments presented against the option; and the implications of the option for the Earth observation sector.

3.2 Free Data for All Users

Definition of the Term

The term 'free of charge' is defined here as no charge to the recipient at the point of delivery. No charge is made for the data themselves, nor for the medium on which the data are distributed. Earth observation data are therefore received by a user without any financial charge to that user.

Arguments in Favour

1. Sharing of environmental data, including Earth observation data, should be encouraged with the lowest possible barriers to this sharing. Data supplied to all users free of charge allows open and easy sharing of these valuable environmental data.
2. A free data policy encourages the widespread use of Earth observation data and hence contributes to the development of the sector.
3. The policy is simple to administer.
4. The tradition of free exchange of meteorological data has been highly successful and suggests that environmental data for environmental research and global change research may benefit from a similar policy.
5. Experience in the United States suggests that the free exchange of basic meteorological data actually contributes to the commercialisation of the Earth observation sector by encouraging the value added sector to develop applications on the basis of free data.

Arguments Against

1. If large volumes of data are available free of charge, there may be no discipline in the demands by users for the data.
2. If no cost is associated with the data, they may also be perceived to have no value.

3. The supplier continues to pay for the data rather than the user, so the user does not have a sufficiently clear say in the amount and type of data collected.
4. Free supply has not generated sufficient recognition of the value and economic impact of the data, particularly meteorological satellite data. This makes it more difficult to justify increased budgets for satellite programmes.
5. If data were supplied free to all users the costs of providing them would need to be recovered through public and/or private sector funding, which would not necessarily be in proportion to the use of the data made by different contributors.
6. The more successful an Earth observation programme, the greater the costs falling on the data supplier rather than on the data user.
7. Some users are able to make a commercial gain from the use of the data without making a commensurate contribution to the costs.

Implications of Free Data for All Users

Providing data free of charge to all users maintains the cost of providing the data on the supplier. In a closed loop system such as EUMETSAT this is acceptable because the community which funds the system is also the recipient of the data. When EUMETSAT wishes to enhance its capability, for example by a new Meteosat satellite, it requests its member agencies for further financial support.

In an open loop system the provision of data free of charge limits change or development of the supplier organisation. In the case of ESA the supply of data free of charge would mean that the funds which member states provide to ESA are locked into an inflexible framework whose justification is governed by its mode of operation; that is, in order to continue to provide data to users, the framework of supplier pays must continue.

Data supplied free of charge to all users could be perceived initially by users to be a positive initiative. However, the cost of data is only one part of the total costs of a research, operational or commercial programme. So, by seeing the data as free, users may underestimate the other, supporting tasks which are required for the successful completion of a programme.

It is not the cost of data that curtails use by the research community but rather the availability of research funding. If funding is not available for research using Earth observation data, including labour, equipment, information and overhead costs, then free data will have little more than pictorial or curiosity value.

3.2 Marginal Cost Price to All Users

Definition of the Term

The price which recovers the costs incurred in providing data beyond the costs of the basic ground infrastructure is the marginal cost price. This is the short-run marginal cost price because it does not seek to recover the ground infrastructure investment. A broader definition of a long-run marginal cost price would encompass the funding necessary to modify and develop the ground infrastructure in response to changing needs.

The marginal cost price is impossible to define accurately, partly because it is the result of administrative decisions, and partly because organisational accounting data are normally not available in sufficient detail to determine a realistic price that accurately reflects the short-run marginal cost of providing a given data set. Therefore, terms such as the marginal cost of reproduction and delivery and the marginal cost of filling a specific user request have been employed in a general sense to give supplier organisations flexibility in how they treat the definition of a marginal cost price.

Arguments in Favour

1. Making all users pay for data, even a nominal price, encourages discipline in the selection of and requests for Earth observation data.
2. The recovery of marginal costs avoids budget deficits resulting from satisfying extra demands for data.
3. The marginal cost price can be set at a sufficiently low level to encourage widespread use of the data.
4. Marginal cost prices allow wide-ranging and flexible demonstrations of applications in all sectors, some of which may later become operational uses. This allows exploration of the data for uses which the Earth observation instrument may not have been intended (e.g. NOAA AVHRR data for vegetation applications).

Arguments Against

1. As in the case of providing data free of charge, the bulk of the costs of providing data need to be recovered by the data supplier.
2. Marginal cost prices do not encourage the eventual transition to a commercial basis for Earth observation systems currently funded by governments.
3. Some users can make commercial gains from the use of the data without making a commensurate contribution to the costs.

4. The costs of administering a marginal cost price policy could be disproportionately high in comparison with the revenues generated. (Indeed, the US policy on data management for global change research states that for small data sets and those data accessed infrequently, the administrative burden of marginal cost recovery may outweigh the benefits of charging such costs, and data may be more efficiently provided at no cost.)

Implications of Marginal Cost Price for All Users

A marginal cost price serves to maintain the status quo. Users receive data at low cost and so any increase in costs at a later stage, such as to commercial prices, will present a structural funding problem to users.

Suppliers do not generate any additional income from marginal cost prices to reinvest in the provision of new services and products. Such innovations have to be funded from capital programmes which require new justifications. Where new products or services are required by only small numbers of users then such justifications by suppliers will not be easy to carry through.

The maintenance of the status quo means that there is a barrier to evolution in pricing policy. Such evolution is necessary to respond to the changing forces acting on the participants in the sector, such as the changing use of Earth observation for environmental purposes.

3.4 Market-driven or Realisable Prices for All Users

Definition of the Term

Because no supplier has so far recovered the costs of the total space segment and ground segment, the term 'market driven' cannot be used in a full commercial sense. An administrative decision has to be made to identify which components of the data supply chain are included in defining a market-driven price.

An alternative term is the 'realisable price', which is the price that a supplier can obtain in the market for the data or products. In the present state of the Earth observation sector, the realisable price is also that price agreed between a willing seller and a willing buyer. This willingness is not governed by perfect market conditions because of government support for the space segment and part of the ground segment.

Although the term has pitfalls, market driven can be used to indicate the price level, which is strongly influenced by (1) what the market is prepared to pay and (2) which operating costs have to be covered by a supplier organisation. The new, very high resolution systems are converging on a price of approximately US$10 per square km for raw data: this figure may set a benchmark for market-driven prices.

Arguments in Favour

1. This policy helps to stimulate the development of a commercial Earth observation data market.
2. Market-driven prices generate a margin that can be reinvested in the enterprise; marginal prices do not do this.
3. The policy is viable even for research users, provided that research programmes include the price of data as part of their research budgets.
4. For research users, the policy puts the provision of Earth observation data on the same basis as that of other goods and services, e.g. computing facilities, telecommunications services, heating, lighting and rent.
5. The focus is on the decisions of the user rather than on the decisions of the supplier organisation.

Arguments Against

1. A commercial pricing policy can restrict the use of Earth observation data by making them too expensive. This can particularly restrict access to the data by researchers, which in turn slows down the rate of progress in Earth observation, and hence the rate at which some applications can reach maturity, to the long-term detriment of the sector as a whole.
2. Pricing policies defined by the space agencies are concerned with data rather than the information required by end-users. A market-driven pricing policy can make the acquisition of information derived from long-term, global data sets very expensive.
3. As most Earth observation data are eventually used for government purposes at present, governments should not have the extra financial burden of market overheads in obtaining data.
4. The policy is likely to restrict the use of data by less developed countries on cost grounds.

Implications of Market-driven or Realisable Prices for All Users

Market-driven prices in principle give power to users. When users pay for data and products they have a strong voice in the definition of those data and products. This is the case with EUMETSAT and the European national meteorological services.

The danger is that research users will have insufficient financial resources to pay market-driven prices. This means that the research use will be underdeveloped, and in turn the operational and commercial uses will also be underdeveloped because they draw upon the research sector for many innovations and new developments.

An increase from marginal costs to market-driven prices has been blamed for the significant decline of Landsat data use by the academic community in the United States (Shaffer 1992). Figure 6.3 shows the number of Landsat items purchased by the academic sector in financial years 1973–1990. The data in Figure 6.3 include both photographic prints and digital data.

The peak of Landsat purchases by the academic sector was in the mid-1970s at a level of the order of 20 000 items per annum. This peak coincided with the Large Area Crop Inventory Experiment (LACIE). During the period of NASA's university grants programme (1973–1982) there were over 8000 Landsat items purchased by the academic sector each year. In 1983, NOAA introduced commercial prices for Landsat digital products, and in 1985 the Landsat commercialisation contract began. The commercial, or market-driven, prices were associated with a decrease in the number of Landsat items purchased by the academic sector: down to 2000 items in 1985 and 450 in 1990.

However, the decline of Landsat sales to the academic sector began in the late 1970s, and from the curve in Figure 6.3 it can be seen that the NOAA and EOSAT increases in prices had no discernible effect on the downward progress in the level of demand by the US academic sector. Therefore, on the basis of this evidence, the simple relationship of an increase in price (to market-driven levels) resulting in a steep decline of demand by the research community is not borne out by the Landsat experience in the United States.

What appears to be happening is that funding programmes that used Earth observation were in decline and the Earth observation research community therefore had less funding for research, including less funding for purchasing Earth observation data. In addition, alternatives to Landsat data, such as NOAA AVHRR, were also becoming available to the research community.

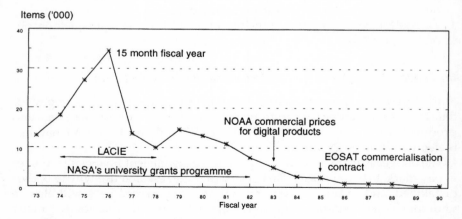

Figure 6.3 Landsat Items (Photographic Prints and Digital Scenes) Purchased by the US Academic Sector, FY 1973–1990. Source: Silvestrini (1991)

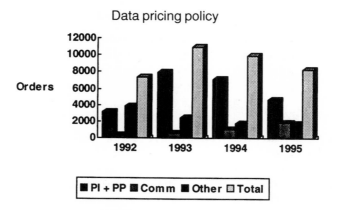

Figure 6.4 Trends in Orders for ERS SAR Data, 1992–1995. PI: Principal Investigator Orders; PP: Pilot Project Orders; Comm: Commercial Sales. Sources: ESA (1996b) and Jensen (1995)

A similar situation can be seen with orders for ERS SAR data. The orders for ERS SAR data for the period 1992–1995 are shown in Figure 6.4. The data show that the orders are dominated by those from the Principal Investigators and Pilot Projects. These projects, which were approved by peer review following an Announcement of Opportunity call for proposals, are supported by ESA through the provision of data at no charge – a form of grant to the researchers. When this support is provided, comparable with the grants available for the use of Landsat data, then use of the data is higher than would otherwise be the case. The examples of Landsat and ERS support the case that the issue is not simply the price of Earth observation data but the structure in which the data are being used.

Because market-driven prices are not full commercial prices in the sense that they do not recover all costs, there is an arbitrary decision to be made about the level of the market price. This decision on pricing level need not be singular. There are benefits in a staged transition from (say) marginal cost prices to market-driven prices. NOAA has increased its environmental product prices to commercial companies but has done so in a way that enables those companies to generate revenues from their own customers without producing a shock to their economic system and their market.

3.5 Full Price

Definition of the Term

Prices set at the full price, or competitive market price (Ordnance Survey 1996), would capture the investment costs of building and launching an Earth observation satellite and its payload, plus the costs of the ground segment and the marketing activities. In addition, the price would reflect the profit needed on any one satellite and the investment in research and development for the next generation of Earth observation satellites.

Arguments in Favour

1. A full, commercial price would recover all the initial investment costs.
2. It would be possible to investigate easily the development of new instruments directly geared to user needs.
3. Full prices can provide a basis for the sustainable and long-term growth of the Earth observation sector independent of government funding.

Arguments Against

1. The price levels would be very high. As an example, what would be the data product cost for an Envisat ASAR product under a full price policy? The cost of Envisat is approximately £1.5 billion. If ASAR product sales of 20 000 per year could be achieved then over the five-year life of the satellite the price per product would be of the order of £15 000. The estimate of 20 000 product sales per year would be twice the current ERS SAR product orders per year and ten times the current commercial SAR orders. At a price of £15 000 per product, the value of using the product would have to be high to justify this level of investment by the buyer.
2. A full price approach would bring a step change in pricing policy for which the Earth observation sector is not ready at present.
3. A full price policy would also fail to recognise the need to invest in space for scientific and humanitarian returns as well as for operational and commercial benefits.

Implications of Full Prices

Many governments, particularly those in Europe, see an important goal in producing an Earth observation sector that eventually has a commercial basis. For example, the 1996 Forward Plan of UK space policy produced by the British National Space Centre has as one of its Earth observation objectives (BNSC 1996) to 'Create the conditions under which a commercial industry in EO, competitive in the world market and fully sustained from private and public sector operational users, can be created by 2005.'

In the longer term this commercial basis is both desirable and necessary because it cannot be anticipated that governments will continue to fund Earth observation indefinitely. The key question is when will this transition occur?

Data pricing policy   119

There is already some evidence of the transition happening with the very high resolution systems initiated by US companies following the deregulation of what was formerly defence technology. In the 21st century, ESA plans to launch satellites of the Earth Observer type which will need to have some element of full prices, even if the full price is wrapped as a contribution to the build and launch costs, as is the case with EUMETSAT.

3.6 Two-tier Pricing

Definition of the Term

A two-tier pricing policy has been proposed by several organisations (e.g. ESA and Radarsat International) to allow preferential treatment in financial terms of research (and other) users. The two tiers of pricing are normally market-driven prices for all users, except for the users in the categories given preferential treatment who receive data at some lower price. The lower price can be one of three levels:

- No charge for the data or for the medium on which the data are delivered.
- Marginal cost price to cover the actual extra cost of providing a given data set (see the discussion of marginal cost pricing above).
- Discounted market-driven price. In principle this is any price (including no charge) below the market-driven price which a supplier wishes to charge a user. In practice the price level is normally somewhere between the market-driven price and the marginal cost price. Discounted market-driven prices are used by suppliers (e.g. EURIMAGE and SPOT Image) to promote certain uses which are beneficial to the supplier organisation.

Arguments in Favour

1. The research community, through publications and other channels, is contributing to the public welfare. Therefore, data for research use should be available for a small charge to cover marginal costs.
2. At the same time, commercial users should pay a price that reflects the value which they derive from the use of the data.
3. The policy gives flexibility to adapt to changing circumstances. For example, if research use develops into commercial use over the long term, then a two-tier policy can accommodate this development.

Arguments Against

1. If Earth observation data are to be widely used, markets must be stimulated and hence any new policies should not undermine commercial markets. This policy could seriously do so.

2. A two-tier pricing policy requires considerable policing to ensure no abuses of the system occur and could, therefore, be unsustainable. The borders between scientific research in Earth observation and practical, commercial applications are difficult to define and short, quick transitions can be made between the two.
3. A two-tier policy will add administrative costs overall which must be borne somewhere in the data supply chain.

Implications of Two-tier Pricing

A two-tier pricing policy is an arbitrary structure, and consequently it can create tensions between the categories of users included in the two tiers. Research users who qualify for the lower price tier may leak the data thereby purchased into a use which falls in the higher tier category. This means that a significant element of a two-tier pricing structure is the monitoring and policing of the implementation of the two tiers.

NASA is in a powerful position to police its two-tier structure because it has a closed loop system and can cut off data and grants to those users who violate the conditions of the lower price tier. However, the experience of NOAA is that it is very difficult in practice to police a two-tier structure, particularly when copying and distribution of Earth observation data are so cheap and easy. NOAA's fees for environmental data are subject to US law 1534 (United States 1990) which states a two-tier approach to environmental product pricing. The core of US law 1534 is as follows:

Except as otherwise provided in this section, the Secretary [of Commerce] is authorized to assess fees, based on fair market value, for access to environmental data and information and products derived therefrom collected and/or archived by the National Oceanic and Atmospheric Administration.

The basis of the fee level is the fair market value of the data and the information and products derived from the data. In principle, the fair market value is the price that would be agreed upon by a willing seller and a willing buyer. NOAA (1991) argues that this simple legal concept cannot be applied to its data and products because of its monopoly position in the market, its unwillingness to recover all of its investment costs in environmental data gathering and the condition imposed by US public law 101-508 that increases in revenues to the United States resulting from increased fees for environmental data should be limited to $2 million in each of fiscal years (FY) 1991, 1992 and 1993, and $3 million in each of fiscal years 1994 and 1995. Therefore, NOAA's fair market value is what is adjudged reasonable based on historical fee levels.

NOAA charges less than the fair market value under the following circumstances:

The Secretary [of Commerce] shall provide data, information, and products . . . to Federal, State, and local government agencies, to universities, and to other nonprofit institutions at the cost of reproduction and transmission, if such data, information, and products are to be used for research and not for commercial purposes.

NOAA's two-tier pricing policy is therefore founded on public law. Despite this legal protection, NOAA still finds it very difficult to police user compliance with the rules, and relies on the threat of prosecution and public disgrace as the main enforcement mechanism.

The implication of the NOAA experience is that while there are benefits in a two-tier pricing policy, there are significant problems of policing which act against a successful use of this approach. It is important to note that other organisations will have a different view of the issues of fair prices and so the responses to the two-tier approach will also be different.

3.7 Information Content Pricing

Definition of the Term

Pricing policy in Earth observation has largely been designed around selling scenes of data. SPOT, Landsat and ERS data, for example, are sold by scene or by parts of scenes. An alternative model, different in character, is to sell Earth observation data by their information content rather than by their geographical extent.

An example of selling data by information content is the monitoring of oil slicks in SAR data (see Pavlakis, Sieber and Alexandry (1996) for examples of oil slick detection in the Mediterranean). It can be envisaged that a customer would wish to purchase only those SAR data in which evidence of an oil slick is present. This in turn requires an assessment of the information content of the SAR image to examine the data for the presence of an oil slick. Only if a suspected oil slick were present would the customer purchase the data, and maybe then only the information on the location of the suspected oil slick rather than buying all the SAR data.

A different architecture of the ground segment may be required to make such a policy successful (Cudlip *et al* 1996). The processors which assess the information content of the data, sometimes called 'sniffers' or 'watchers', could be located at the customer site, although in the case of SAR images which contain no oil slicks then the transfer of the data to the customer site would be nugatory.

Alternatively, the processors could be located at the data reception station and perform the assessment of information content in near real time as the data are being received. Only when the processor identifies an oil slick or other required information would the data or derived information be transferred to the customer site. A pricing policy based on information content could therefore place an extra requirement for data processing at the data reception site.

Arguments in Favour

1. The focus of the approach is on information rather than just data.
2. The customer would only buy data (or information) that was directly relevant to the application.
3. The data delivery system would be geared to an information service and to value added information related to customer needs.
4. There is greater scope for the development of data brokers and other contributors to the Earth observation value chain.
5. The price would depend on the quantity of relevant information rather than on the volume of raw data.

Arguments Against

1. Because the focus is on the information content rather than the volume of data, this leads to questions of the value of the information. In turn, the value of the information depends on the organisation using the information and the role of the information in the organisation's value chain.
2. The disadvantage of an information content pricing policy revolves around how to set the prices. For the example of oil slicks the question becomes one of how much the information is worth to the customer. This in turn depends on the customer's ability to pay and an assessment of the value added by the extra information on the location of oil slicks.

Implications of Information Content Pricing

There is a class of events for which information content pricing is a useful approach. For episodic events such as oil slicks, earthquakes, floods, major storms and volcanic eruptions, an information content pricing policy can be realistically envisaged.

For scientific and operational applications that require data on a regular basis, such as crop monitoring, sea surface temperature mapping and ice edge detection, the case for setting a price based on information content is not so clear. The value that a customer gains from, say, crop yield predictions using Earth observation data very much depends upon the responsibility and size of the customer organisation and the role which crop yields play in that organisation. For the European Agriculture Guidance and Guarantee Fund, for example, crop yield predictions influence every year the market-making activities such as export restitution. For farmers, crop yield predictions have value in some years and little value in others, depending upon market and crop conditions.

3.8 Rebalancing of Government Funding

Definition of the Term

One issue that has emerged from the data pricing debate is the extent to which preferential treatment should be given to certain users. A widely held view is that data to be used for scientific research, and possibly also for public operational services, should be supplied at prices which are lower than the market-driven (realisable) rates to be charged in other cases.

An alternative view is that a particular data product should be supplied at the same price irrespective of the intended use or type of user. Proponents of this second view recognise that research users (and possibly some operational users) have limited funds for the purchase of data, but they believe that assistance to such users should be provided through mechanisms other than differential pricing.

Behind these views of how to treat certain categories of use or user is a concern for two key issues:

- to ensure that data are diffused widely in support of the socially valued activities of research and operational uses for the public benefit;
- to create transparency in the market so that the origin of resources is clear and may be more closely matched to the destination of resources.

One pricing model that can respond to these concerns is a rebalancing of government funding (Mansell, Paltridge and Hawkins 1992), and this model is discussed below. The rebalancing is particularly set against a two-tier pricing policy.

In the model of rebalancing government funding, space agencies and data intermediaries would receive realisable market prices from all users. Governments would then provide financial support for research and public-good users at a level that would match reductions in assistance to space agencies and data intermediaries. Under the rebalancing model the flow of funds from government would go not to the providers of the data but to the research and operational users of the data, who would then purchase data from the provider organisations at market rates. In both cases the other users are not affected by any change.

The key policy change that rebalancing implies is a shift in financial resources from the space agencies and data intermediaries to users whose activities are acknowledged as being in the public interest and are funded by the tax-payer. In a perfectly competitive market this change would not increase the financial burden on government (transaction costs aside). However, it may mean transactions internal to governments on the origin of financial resources to establish a fund for research and operational users.

Arguments in Favour

1. The user pays, and so the services provided by the data and product providers are more closely geared to the needs of the user.
2. The competition produced amongst data providers would make them plan their activities explicitly towards their user base.
3. Government assistance to public-good users becomes more transparent.
4. Funding sources that are explicitly recognised and guaranteed by government would overcome fears in the research community that governments are not prepared to provide ongoing support.

Arguments Against

1. A change would be necessary in the pathways of government funding and this may be difficult to achieve.
2. The category of research and operational users which receives direct government support has to be clearly and unambiguously defined.
3. Leakage to users not in the research and operational category (e.g. commercial consultancy) may still occur and would require vigilant monitoring.
4. Researchers not qualifying for funding, on geographical or other grounds, could access data free of charge or at marginal cost prices from other sources and so hold back the overall growth of the Earth observation market.
5. Data and product prices might rise with greater reliance on the market mechanism and less reliance on guaranteed financial support from governments.

Implications of Rebalancing Government Funding

While Earth observation is in principle a public good, it is not a simple public-good information market because of the distortions brought about by the combination of public and private sector decisions.

The rebalancing model would entail movement towards greater reliance on market-driven prices and there would be a need to take account of the ingredients that markets need to function efficiently. These include ease of market entry and exit, likelihood of competitive pricing and the widespread availability of market information.

The rebalancing model would require sources of government financial resources to be explicitly established. The approach is comparable to a proposal of Macaulay (1992) for the introduction of information vouchers: 'These [information] vouchers would take the form of data access grants financed by government entities. The grants would be awarded to scientists, nonprofit organizations, and others deemed to be acting in the public interest in using remote-sensing data.'

As a first step in the rebalancing of government funding it would be possible to build simple administrative structures to identify the planned budget available for data to be provided to principal investigators in the scientific sector. The principal investigator would pay for data using his/her information vouchers. This would give a notional budget to the approved scientific community where actual data costs are zero but accounted costs are set at the market-driven (realisable price) rate.

The focus of the rebalancing model is on the user. In order for the rebalancing model to provide a good match with policy objectives, researchers would need to have the same rights as commercial users to choose their sources of data. The European Science Foundation strategy for Earth observation (ESF 1992), for example, notes: 'The user should have choice of access to alternative products provided by different facilities competing for support on the basis of proven performance.'

Linking research grants to particular data suppliers would reduce the effectiveness of the market mechanism because space agencies and data intermediaries would not have an incentive to increase customer responsiveness and might increase prices to captured customers.

4 CONCLUSIONS

Earth observation data pricing policy is a complex issue because there are many differing perspectives on the problem. There is a need for greater clarity in the objectives of Earth observation missions and how these are tied into pricing policies. In that sense pricing policy should be the servant of the mission objectives and not the other way round.

There is an increasing emphasis on meeting the needs of users and so gearing the data capture, data supply and pricing policy chain to specific user needs.

The speed and nature of the transition from public funding to a greater supply of funding by users is a delicate issue and one that can have significant implications for the development and maturity of Earth observation.

7
Data Preservation

1 INTRODUCTION

1.1 Illustrations

Earth observation data constitute a resource that requires significant investments to capture. A key question is what happens to these data in the long term? The definition of 'long term' is not clear. Is it 5 years, 50 years or 500 years? At present, the policy questions of long-term data archiving are largely unasked, and data archiving is scheduled to fit in with other activities of the data supplier organisations.

This introductory section selects three examples to show the concern for Earth observation data archiving voiced by several organisations.

1.2 Columbus Utilisation

During the 1980s the British National Space Centre prepared for the next generation of Earth observation systems, which at that time was envisaged as the international Space Station Columbus project, by carrying out a series of user-requirement reviews which addressed thematic topics. The Columbus land science and applications review proposed that there should be a dialogue between applications centres, data centres and data suppliers on both data capture and data disposal strategies. The review proposed the following three levels of archive:

- primary data to be archived for 3–5 years
- secondary data to be archived for 5–10 years
- tertiary data (the very 'best' data) to be archived permanently

1.3 UNESCO

At the European International Space Year conference held in Munich in March/April 1992, the United Nations Educational, Scientific and Cultural Organisation (UNESCO) presented a paper on the subject of the protection of satellite Earth observation data from the viewpoint of the data user (Kumon 1992). The paper makes a clear statement on the need for long-term data protection:

In view of the global change researches, the researchers in 21st or 22nd century should be guaranteed with the access to data of the 20th century.

However, it is said that the satellite operators policy for data conservation is for 10 years and earth observation data acquired during 1970s are deteriorating in their archives. Therefore urgent countermeasures should be taken for data conservation.

1.4 US National Archive

During the 1990s there has been a liberalisation of US policy concerning Earth observation. The Land Remote Sensing Policy Act of 1992 (US Public Law 102-555), which provided a new framework for the Landsat programme, and the remote sensing policy signed by President Clinton on 10 March 1994 both increased the flexibility for operators of Earth observation systems and services. This liberalisation did carry with it some constraints, and those constraints from the 1994 presidential directive relevant to data archiving are as follows:

An operator shall make available unenhanced data requested by the National Satellite Land Remote Sensing Data Archive of the Department of the Interior.

The Archive may make these unenhanced data available to the public after a reasonable period of time, this period of time to be agreed with the operator.

Before purging any data in its possession, the operator must offer the data to the Archive. The Archive may subsequently make the data available to the public.

1.5 Nature of the Problem

These three illustrations raise many of the issues involved in the data archiving question. There are two central concerns: for how long should Earth observation data be archived and which organisations should take responsibility for the archives?

Data suppliers such as ESA typically have a commitment to preserve their Earth observation data for approximately 10 years after the end of the mission. This means effectively that they have a budget line to support the data archive for a mission for 10 years beyond the completion of the mission: this is the case for the ERS-1 and ERS-2 missions. SPOT Image have so far kept all the SPOT data, even though the data above three years old have little sales value, because the cost of selective archiving is greater than the cost of archiving all the data.

When we look to environmental issues the need for access to archive data is often greater than 10 years. For example, river basin planning has mainly been undertaken using hydrological records of 15–30 years in length and this has proved inadequate for the robust design of dams and other water management systems. The River Nile had a particularly long hydrological record of some 60 years when the Aswan High Dam was designed, but even this has proved inadequate for Nile basin management. River basin planning requires 100–200 years of environmental data to produce reliable information on basin changes and behaviour.

When we look at climate change issues (Houghton *et al* 1995) the need for data records is comparably long, with data archives as long as 150–200 years needed to examine some of the impacts of climate change.

There is therefore an imbalance between the archive plans of the data suppliers and the archive expectations of at least some of the user communities. Which organisations will take responsibility for long-term Earth observation data archives that will be useful in the future for research and commercial applications?

2 MODES OF WORKING

2.1 Technology

There is a technology issue that is important in the archiving question. Until recently much Earth observation data was archived on magnetic tapes. Tapes have a limited life and the data on them need to be rewritten at least once every 10 years and preferably once every five years. Rewriting once every 10 years still carries an expected loss of data from 1 in 100 tapes. In addition, tapes are prone to physical damage either manually or by tape-reading equipment.

Data suppliers have found that magnetic tapes are not a reliable medium for long-term storage because of technology obsolescence. NASA saved only 20% of Landsat 1, 2 and 3 data because the tape-reading equipment became obsolete. A similar problem occurred at ESA where only 2000 of the 15 000 tapes containing Landsat data were saved before the tape-reading technology became unusable.

Fortunately, the current storage media of CD-ROMs and optical discs have much longer lifetimes than tapes. However, much valuable Earth observation data are still held on tapes and while their potential value may increase in the future in the context of environmental change programmes, their physical medium will inevitably deteriorate.

Data storage costs are likely to decrease by an order of magnitude every 4–5 years. This means that by the year 2005 it will cost only about 1% of the cost in the mid-1990s to store the same amount of Earth observation data.

2.2 Distributed Access

The early uses of Earth observation data were characterised by single applications on single data sets. The number and range of applications are increasing, and the number of data sets used either separately or in conjunction is also increasing. The global Internet is assisting the rapid access to a variety of data sets. Figure 7.1 shows how a user's desk-top interface increasingly controls a wide variety of ways of using a variety of data, and shows the importance of archive data that are accessible and easily usable. In fact, if the Earth observation data archives are not easily usable then this will be a significant bottleneck in the use of Earth observation data.

In the next 10 years the use of distributed archives will grow following initiatives such as the Distributed Active Archive Centers (DAACs) in the US EOSDIS and the Centre for Earth Observation (CEO) and the European Wide Service Exchange (EWSE) in Europe. Access to a distributed data archive system will be enabled more easily by increased bandwidth available from commercial communication carriers. In the mid-1990s a common communication link can deliver 64 kbits s^{-1}. This is likely to improve to 10 Mbits s^{-1} by the year 2000, and to 1 Gbit s^{-1} by the year 2010. This will improve data transfer rates so that while it takes 18 hours to transmit a full ERS SAR image at 64 kbits s^{-1}, by the year 2010 it will only take 5 seconds with a bandwidth of 1 Gbit s^{-1}.

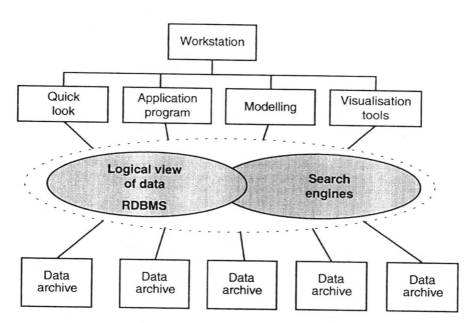

Figure 7.1 Interactions of Users with Data Archives. Users Interact with the Relational Database Management System (RDBMS) and with Search Engines Rather Than With the Data Directly

Table 7.1 Metadata Requirements. Source: NRC (1991)

- Background information
- Data collection procedures
- Data transport and verification statistics
- Definition of calculated variables
- Definitions of calibrations applied
- Description of adjustments
- Description of instrumentation
- Full or sample listings
- Full variable definitions
- Identification of contributors
- Input/output routines on the transport medium
- Limitations of the data
- Modifications made at the processing centre
- Quality assurance at the centre
- References
- Scope and purpose of the programme
- Station history
- Systematic and random errors

3 METADATA

Earth observation data are a representation of reality. In the process of measuring the Earth, the representation of an environmental variable changes at each stage between electromagnetic energy entering a sensor and the final product being produced. The measurements made in the space segment are a model of the actual environment, while in the ground segment the model is inverted to retrieve an estimate of the original environmental variable. After the inversion process in the ground segment, other processes are employed which affect the characteristics of the data, e.g. truncation and resampling. At each stage of this process there is a potential loss of information.

It is therefore vital at each processing or transfer stage to characterise the quality of the data, so that the user is aware of the processes occurring at each stage in the chain. The quality of the data refers here to an accurate description of the data or product: it has also been termed the pedigree, the audit trail of the data or the metadata.

Metadata are those data and information that describe the original data set. In a simple form the metadata will include the date and time the data were collected plus the spatial coordinates of the location of the data. SPOT products at levels 1A and 1B include these data plus the values of the calibration coefficients from the CCDs on the SPOT HRV instrument.

Increasingly, this simple form is seen to be inadequate, and more and more information is included in the metadata. The US National Research Council has produced a strategy for managing data and information in the context of global change research, and in their strategy they noted the importance of the '20-year test' (NRC 1991): 'will someone 20 years from now, not familiar with the data or how they were obtained, be able to find data sets of interest and then fully understand and use the data solely with the aid of the documenta-

Data preservation

tion archived with the dataset?' For documentation read the broader topic of metadata because documentation of data sets is a specific case of metadata. The NRC (1991) goes on to list the minimum metadata requirements shown in Table 7.1.

The authors of the NRC report also note that: 'Complete data documentation is a crucial portion of data processing and, along with quality assurance, will be the heart of successful data management activities for global change.'

It is arguable that only raw data should be archived, and no higher level products archived, as long as the algorithms and the supporting data to produce the higher level products are also always archived in the metadata. For simple operations such as geometric correction of SPOT or Landsat data this seems a straightforward procedure. However, for data sets that rely on calibration campaigns using *in situ* or airborne sensors then there is a question of how extensive the metadata should be. In such circumstances should all the supporting data be archived as well? If so, this makes the cataloguing and access tasks more difficult to achieve smoothly.

As the growth in the quantity and diversity of metadata continues, there is an argument that all data should be treated alike and there should not be a distinction between data and metadata: one man's data is another's metadata. The distinction between metadata and data may be useful conceptually, but in systems to process and store Earth observation data it is becoming less useful to create systems that have hard boundaries between these two data types.

The tools to interrogate Earth observation databases are important. CEOS has produced recommendations on the categories of information services and suggests the four different levels of information listed below (Triebnig 1994, 1995):

- *Directory*: brief descriptions of whole data sets (e.g. ERS-1 SAR data) suitable for identifying data and data products.
- *Guide*: detailed information suitable for assessing the application value of whole data sets.
- *Inventory*: attributes of what are called data granules. Data granules refer to a scene of a group of data that have an identifiable location and date. This level of detail is required for product ordering.
- *Browse*: visualisations of individual data granules, such as SPOT or Landsat quick-look images. Useful for assessing areal extent and data quality.

In addition, there is also progress on how users can reach Earth observation data. There are more tools now available which go beyond simple access to one catalogue. Transparent user access to data held in various archives by interrogating catalogues with common search parameters is becoming easier through tools such as the German Intelligent Satellite data Information System (ISIS) held at the following WWW site:

http://isis.dlr.de/

4 POLICIES AND CRITERIA

While there is a growing volume of Earth observation data and a growing recognition of the need to retain Earth observation data for use in the future, there is no overall, international strategy to manage the process of Earth observation data preservation.

In the United States there is the National Satellite Land Remote Sensing Data Archive, which is operated by the Department of the Interior. This archive has a right to request data which would otherwise be purged by US satellite operators.

Where such archiving agencies are developed, at national or international levels, what are the issues that they should consider? This section presents the main issues that should be considered when archive strategies are being developed.

- *Data types*. Not all data types for all parts of the Earth are required in long-term archives. High priority should be given to data that can later be analysed to examine global change.
- *Levels of data*. It is arguable that, provided there is a good metadata base, it is only necessary to retain low level data, i.e. raw data or level 0.5 data, because the higher level products (levels 1 and above) can be created from the low level products plus the metadata. This imposes greater demands on high-quality metadata.
- *Archive frequency*, which may be related to the nature of the application. It may not be necessary to keep all the SAR data over the oceans, although it would require confirmation from user communities such as the Global Ocean Observing System that this is the case.
- *Duplication of data*. Unnecessary duplication of archives should be avoided or minimised. The improving communications capabilities are providing better access to remote data archives.
- *Archive structure*. It is imperative that the structure of the archives allows easy access to the Earth observation data by users.

At present, most Earth observation data are retained, subject to the performance of electronic media. Because storage costs are in steep decline, the cost of retaining Earth observation data may not be a significant issue, but what is important is the need to manage the data and to enable users to locate and access the data. Ahead of deciding the storage strategy there is also value in deciding on the policies for capturing data in the first place. This means that guidelines for instruments and their different modes of operation should be developed so that data capture is efficient. For example, the Envisat-1 ASAR will have up to 30 different modes of operation: guidelines on which of these modes provide useful long-term data will be useful to the archive managers.

5 CONCLUSION

Most Earth observation data are retained in some form of archive. While storage media are becoming physically smaller and more robust, the task of archive management has not been reduced. Users in the scientific and commercial sectors will continue to require selections of data based on information content rather than on date and location. Policies developed on archiving should therefore take into account how the data are likely to be used in the future as well as how they are provided by the satellite operator.

8
Conclusions and Recommendations

1 CONCLUSIONS

1.1 Conflict

There is conflict in Earth observation data policy. There are different perspectives based on the following:
- the open supply of environmental data funded by the taxpayer and intended for the public good,
- the enthusiasm to recover part of the very large investment in Earth observation, and
- a desire to see a growth of Earth observation independent of government which would follow the precedent set by satellite telecommunications.

In her testimony to the US House of Representatives on Earth observation data policy in 1991, Macaulay noted two attributes of Earth observation that have complicated the design of public policy in this area (Macaulay 1991):

> One attribute is that [Earth observation] data are an information good or commodity, thus having once invested in the infrastructure to obtain the data, multiple copies can be made at typically very low costs of reproduction.

> [The second attribute of Earth observation data] is that the data are not just a source of private information, but are information about public goods – natural resources, climate change, pollution. It is inherently difficult to establish the value of information about public commodities and processes.

The different perspectives on Earth observation data policy have led, and are continuing to lead, to conflicting data policies which are restricting the growth of the Earth observation sector.

1.2 Public and Private Sectors

Earth observation is dominated by the public sector. The main source of funding for Earth observation is governments, acting either independently or

Conclusions and recommendations 135

through international clubs such as ESA. The real data policies that govern access to Earth observation data in practice are developed by the owners of the data. Therefore, it is governments that are effectively responsible for the development of the Earth observation data policies in use today and for much of the near term.

The role of the public sector depends upon whether it is customer or sponsor. If it is a customer for data or information that can be provided by Earth observation then it acts in just the same way as a commercial organisation to procure either the data or the means to acquire the data.

The UK Department of the Environment (DoE) is a good case in point. In collaboration with the UK Natural Environment Research Council and the Australian government, the DoE is procuring the Advanced Along Track Scanning Radiometer (AATSR) which will fly on Envisat-1. The AASTR will collect sea surface temperature data to continue the record of such measurements started with ERS-1 in 1991. The record will thereby be continued until approximately the year 2005. In procuring the AATSR the DoE is acting as a commercial customer for the instrument to fulfil its own objectives.

Another Envisat-1 instrument is the Advanced Synthetic Aperture Radar (ASAR) which is funded by governments as part of the instruments developed by ESA on the Envisat-1 payload. This does not have a dedicated customer who is providing the funding for the instrument, but is part of the process of developing the SAR technology to service a growing user base for the information it can provide.

Here is the contrast between government acting in a customer role and government acting in a sponsorship role. As a customer with a defined information requirement, the government funding route can lead to a sustainable Earth observation sector. As a sponsor with a desire to develop innovative technology, the government funding route is less certain and requires a continuing commitment to this form of funding.

The situation in the United States is for government to act as both customer and sponsor. Ultimately the Mission to Planet Earth has a customer in the US Global Change Research Program and this programme funds both the technology development and the science users to exploit the data. As long as the USGCRP justification is present, and as long as the funds are voted each year, this approach can be sustainable. If there were future government policy changes over support to global change research then the withdrawal of the justification for the USGCRP would result in changes in funding for Mission to Planet Earth. NASA has tried to develop some commercial uses of Earth observation data (CES 1995; Macaulay 1995), although compared to the spend on global change research, the level of commitment is small and its justification uncertain.

Given the dominant role of the public sector, the private sector has mainly acted as a contractor to public sector procurements. This situation is changing with the new US very high resolution systems which are mostly initiated, funded, procured and operated by the private sector, but overall the private sector tends to operate as a contractor in the nations active in Earth observation. There are the following three types of company in the private sector in Earth observation:

- Manufacturing companies which build the space and ground infrastructure, such as Matra Marconi Space Systems, Hughes, Lockheed Martin and Aerospatiale. The companies employ hundreds of staff, have highly specialised spacecraft construction facilities and base their businesses on a succession of large contracts. The companies have an interest in applications because they see the potential for future business through the expansion of the applications market.
- Value added companies use Earth observation data to produce products and services for end-user companies. The entry barriers are relatively low for these companies, and many have started up in home offices by offering Earth observation expertise to niche markets. Staff numbers tend to be in the tens and their markets are diverse in character. The manufacturing companies have invested in the value added companies as one route to market expansion; for example, Matra Marconi Space Systems owns 45% of the National Remote Sensing Centre Ltd.
- System and software engineering companies which develop both value added applications and data processing systems. Earth observation is only one part of the business portfolio of these companies: there may be 50 members of an Earth observation division in a systems engineering company which employs 1000. The long-term business interest of these companies is well served by expansion of the Earth observation applications market because more systems, products and services will be required to serve the needs of end-user clients. Internally the role of staff resources allocated to Earth observation is under constant review because of the need to balance investment decisions in Earth observation against investment decisions in other sectors in which the company operates.

Is this industrial structure adequate to assist in the development of the Earth observation sector, and particularly to enable it to develop a self-sustaining character? The value added companies are too small to have the capital base to do anything other than build their business on data flows from the space data providers, which are typically governments. The system engineering companies are also too small. Given the pressures on their managers to invest in sectors that may be more profitable in the near term, then they too build their business on the space data suppliers. This leaves the manufacturers. With the exception of the very high resolution data providers in the US, this sector appears to be conservative in its approach to the Earth observation market. There is a clear dependence on government as customer or sponsor, which means that the sector may be following rather than leading the development of Earth observation.

There are two elements missing from this description: users and financiers. They are missing because they have been backward at coming forward. The oil companies did invest in their own Earth observation capability in the past, but Earth observation capability is now often contracted out to the value added companies. Other end-users, such as in agriculture or environmental monitoring, also use the value added or system engineering companies as product or service suppliers.

The financial sector seeks an annual internal rate of return of approximately 30% on risk capital for ventures such as Earth observation. This is a high burden for new initiatives and is unlikely to become lower because the infrastructure for Earth observation is remote in orbit. If a financial institution invests (say) £500 million in a road construction project and the project fails, it at least has the road which it can use in a proactive way to attract other investments. If a financial institution invests £500 million in an Earth observation satellite and the satellite fails in orbit, then all it has is a dead and unserviceable satellite, probably backed up by a dedicated ground processing system whose market value is limited to the second-hand sale price of the computers.

The public and the private sectors still work at arm's length as procurers and contractors. It is not surprising that it has proved difficult to develop data policies for Earth observation which satisfy the interests of both of these sectors, never mind the science sector's interests.

1.3 Two-step Approach

The tension that exists in Earth observation data policy is in part caused by the level of maturity of the sector. It has been noted by a working group of UK industrialists and government officials that (BNSC 1995):

A significant obstacle in the progress of satellite Earth observation towards commercial operation is the lack of coherent structure in its market place. The crucial point to overcome this obstacle is to switch the emphasis away from the supplier of satellite data and towards the customer for satellite data.

The working group addressed this problem and proposed what it termed a 'two-step approach' in which commercial opportunities become apparent based on an increasing blend of government action and market forces.

In the first step of the two-step approach, a customer with a substantial operational need for information which can be satisfied by Earth observation expresses a requirement. Public sector capital is used to fund the space and ground infrastructure to meet the information need or, more likely, to guarantee the purchase of Earth observation products and services. A market is thereby created, of limited nature but capable of demonstrating an economic relationship linking buyer and seller. This process indicates how the ingredients of normal market activity, such as market entry and exit, competitive pricing and availability of information, could arise.

In the second step of the two-step approach, private sector investors convinced by the demonstration at the first step would provide the finance for the next generation of satellite systems and seek a broader customer base.

The BNSC has built the two-step approach into its objectives, and it is one of its four strategic actions for the medium term (BNSC 1996). The European Commission has also endorsed the two-step approach, which in this case resulted from a working group of senior European space industrialists (European Commission 1996b).

The first application area for the two-step approach is widely seen to be European agriculture, given the large spend in its support by the European Union each year under the Common Agriculture Policy. There may be other candidates for the first step, such as environment and coastal protection.

What does the two-step approach mean for Earth observation data policy? First, the user is central to the development of the Earth observation sector and exercises this influence by paying for products and services (Smith 1994). Second, to achieve coherent structure in the market-place, the origin of the two-step approach, Earth observation must engage with sources of finance that are customer related and not reliant on government in a sponsorship role. This does not necessarily always mean private sector finance, but it does mean sources of finance that are driven by customers and users and not by technology development. In that sense the comment by the European Association of Remote Sensing Companies (EARSC) to ESA in relation to Envisat-1 data policy is a useful summary. EARSC believes that the driving force for the Envisat-1 data policy should be to foster the development of products and services by European industry and the use by paying customers. This theme can be readily applied to all Earth observation data.

There is danger here. If governments investing in Earth observation were to move tomorrow to a situation where all users paid a price for Earth observation data that reflected the investment costs, there would probably be no Earth observation market because product prices would be too high. The challenge is to achieve that situation in the medium term by a phased transition which acknowledges the legitimate concerns of the scientific, operational and commercial sectors plus the interests of governments in their overall industrial and scientific policies.

1.4 International Diversity

Earth observation is international and becoming more so. Along with the main space players of the United States, Europe and Japan, there is a growing

Conclusions and recommendations 139

importance of India, Brazil, Argentina, Australia and other countries in the launching of Earth observation missions.

Earth observation is also becoming more diverse, particularly with the launches of the US Earth Observing System and ESA's Envisat-1 and Metop-1. The nature of the instruments and their purposes are changing from the collection of environmental information with a broad application base to more specialised information collection to suit particular information needs.

On the user side, there is also a growing internationalisation of Earth observation data use. The major scientific research programmes (IGBP, WCRP and HDP) and the global monitoring programmes (the IGOS components) are becoming greater users of Earth observation data.

There are therefore two contrasting directions of growth. On the supplier side there is a greater diversity of information and because the data come from different organisations there is a diversity of data policies. On the user side there is growth in applications, but there is a greater need for stability in the conditions of access to improve the ease of data availability for these scientific and monitoring programmes. Taken from a user perspective this provides pressure for greater coherence in Earth observation data policies.

One strand evident on Earth observation data policy is a greater level of transparency in the access to Earth observation data. The encryption of Meteosat data, the European Directive on Databases and the US Freedom of Information Act all illustrate the desire to achieve greater clarity in the definitions of access to data. The greater clarity is in turn related to the objectives of the Earth observation missions and the desire to be explicit about the value of the data or the return on the investment. As the French have noted (Synthesis 1995), the return on investment is not only of a financial nature, but can also be measured in terms of environmental security, humanitarian relief or verification of international treaties (Boutros-Ghali 1994). The greater clarity that this approach encourages will assist in improved international understanding of different Earth observation data policies.

1.5 Ground Segment

Data policies have an impact on the architecture of the ground segment. At present, pricing policies are based on an image or geographical area. If the pricing policy were to be based on the information content of a data set, then the processing in the ground segment would have to change.

If there were sniffer algorithms to assess the presence of oil slicks or agricultural fraud in Earth observation data, then the location and nature of that processing would have an influence on the ground segment architecture. Would the sniffer algorithms be located at the customer site or at the data receiving station? Would their location be related to the level of security and confidentiality of the information sought by the user? Which organisation would be responsible for auditing the algorithms: the supplier organisation or the customer organisation? The answers to these questions will influence the distribution of the elements of the ground segment in Earth observation.

Data archiving in the ground segment incurs high media costs. In the future it will incur lower media costs in relation to the volume of data but higher management costs. There are few clear policies on archiving which help to inform the design of Earth observation ground segments. Networking technologies and the Internet are providing opportunities for Earth observation which will allow it to change from being a separate sector to one that can be more easily integrated into other data systems.

2 RECOMMENDATIONS

2.1 Introduction

As noted above, Earth observation is becoming more international and more diverse. What are the issues which those responsible for existing and new missions should consider when developing their statements on Earth observation data policy? The concluding section of this book recommends the positions that should be considered by those responsible for Earth observation data policy development.

The dimensions of Earth observation data policy tend to interlink. It will be helpful for organisations to work on the interlinkages of their policies both externally and internally. While individual policies on pricing, distribution or archiving may be entirely justifiable in their own right, the question must be asked of how they interact together to influence the ways in which the data policy is contributing to meeting the Earth observation mission objectives and to developing the Earth observation sector.

2.2 Mission Objectives

A greater clarity in the statement of mission objectives would help others to understand why the investment in Earth observation is being made. By stating clearly what the Earth observation mission is for, it should be more straightforward to state what the data policy is trying to achieve. Landsat and ERS are examples where commercial intentions have been subsequently added to what were originally designed as public sector missions for scientific and applications research and the flow of a limited amount of operational information.

The primary objectives of the Envisat-1 mission, which are listed below, show how objectives and technology can be intertwined to create uncertainty for data policy (ESA 1994a).

Conclusions and recommendations 141

- To provide for continuity of the observations started with the ERS satellites, including those from radar-based observations.
- To provide for enhancement of the ERS-1 mission, notably the ocean and ice mission.
- To extend the range of parameters observed to meet the need to increase knowledge of the factors determining the environment.
- To make a significant contribution to environmental studies, notably the areas of atmospheric chemistry and ocean studies (including marine biology).

These objectives could be read as entirely in the technology demonstration phase, but in other places ESA calls for a balanced exploitation of Envisat-1 data for both commercial and scientific purposes. Its fourth objective cannot be achieved by ESA itself because it does not engage in environmental studies but is a supplier of data to other organisations and programmes which do engage in environmental studies.

It could readily be argued that these Envisat-1 objectives are produced as a result of European political compromise, and this would be both accurate and acceptable. The key question for this conclusion is: where does this leave the development of Earth observation data policy? Clearer mission objectives will assist both the development of robust data policies and the better understanding in the international community of what those data policies are designed to achieve.

A European Commission report on Earth observation data policy (European Commission 1992b) proposed the following recommendation: 'The pricing policy for Earth observation data should depend upon the class of Earth observation mission. Each class of mission should have a mode of pricing policy for each type of use.' This recommendation of the European Commission implicitly assumes that there are clear mission objectives since it goes on to identify four classes of missions:

- science missions
- research and development missions, which are part of a process of transition to operational missions
- missions declared operational
- missions declared commercial

These missions have different objectives and against these objectives it is possible to identify different pricing models which help to meet the objectives of the mission. The three pricing models proposed by the European Commission working group are as follows:

- supply at not more than the marginal cost of reproduction and delivery;
- supply for all categories of use at market prices, with the exception of certain defined uses, including research uses, which may receive privileged prices;

- supply for all categories of use at the same price, with government funding redirected to the user segment in the interests of producing the most useful data, and maximising their use.

This approach would be more explicit and would help to give greater clarity to what the mission is trying to achieve and how the pricing policy will assist the process of achieving the objectives.

2.3 Proactive Tool

Earth observation data policy should be used as a proactive tool to create better conditions for the development of the Earth observation sector. It can be used, as in the two-step approach, to help the development of operational systems instead of one-off missions.

As the major global scientific and monitoring programmes develop, it is helpful if their data requirements become more formally expressed. This will help to define the user requirements to which the data suppliers can respond. This already happens with NASA because in its Mission to Planet Earth it is both the supplier and the customer. For other programmes an explicit definition of real user requirements will help to define workable data policies rather than simply agreement to international policies, such as those of CEOS, which may not be fully implemented in individual cases.

The two-step approach encompasses explicit user requirements at the first and subsequent steps, and these can be used to drive the development of policies on data access which contribute to the long-term and sustainable growth of the Earth observation sector.

2.4 Explicit Resources

Much Earth observation data are provided for free or at marginal cost, particularly for scientific use. While this procedure will continue for some time, it will be useful to be able to identify the financial value of the data supplied. This may mean using information vouchers and setting up some internal accounting procedures to reflect the financial value of the data which have been provided at a lower price. When data are provided free or at the marginal cost, then the accounting information will allow both the supplier and the user to acknowledge the full level of the transaction.

Earth observation data typically comprise 5–10% of the costs of a project using data. Therefore it is necessary to set the pricing policy in the context of the overall expenditure. For example, up to the end of 1996 EUMETSAT provided some Meteosat data free of charge to educational users. While the data were free there was still an overhead in EUMETSAT staff time and computer resources required to make the data available with no charge for

Conclusions and recommendations

the data themselves; likewise on the side of the recipient user, staff time and computer resources are also required to make effective use of the Meteosat data.

Greater clarity on who funds what will also contribute to improvements in the ways in which Earth observation data policies contribute to the achievement of the objectives of Earth observation missions.

2.5 Archives

Through its Centre for Earth Observation, the European Commission is studying the issue of long-term archives of Earth observation data. There is a US policy on long-term archives of data which are not to be retained by a private sector organisation.

A key question for all Earth observation missions is: who will be responsible for Earth observation data archives in 200 years time? The most likely answer is the public sector in some form, as the guardians of information about planet Earth on behalf of the public. Even if the public sector contracts out the task to the private sector, the public sector needs to establish mechanisms to fund long runs of Earth observation data so that they are accessible to future generations. The opportunity to exploit the investment in Earth observation will be enhanced if, at their outset, all Earth observation missions have an accompanying policy statement on the plan for long-term archiving of the data generated by the mission.

Supplier organisations should monitor the developments of networking and archiving technologies and change as the modal technologies change. It will be possible to make more Earth observation data available on-line, which should provide further opportunities to maximise the use of the data.

3 MAXIMISING VALUE

Many organisations in the Earth observation sector have agreed that maximising the value of Earth observation data by maximising its beneficial use is a highly desirable objective. Data policies have as much a part to play in the process of maximising value as do the development and operation of Earth observation instruments and processing algorithms. Improving data policies will be important for the Earth observation sector for many years, and maybe for as long as it is recognisable as a sector.

Appendix
Members of the Committee on Earth Observation Satellites

Source: CEOS 1995

CEOS MEMBERS

Australia	Commonwealth Scientific and Industrial Research Organisation (CSIRO)
Brazil	Instituto Nacional de Pesquisas Espaciais (INPE)
Canada	Canadian Space Agency (CSA)
China	Chinese Academy of Space Technology (CAST)
	National Remote Sensing Centre of China (NRSCC)
Europe	European Commission (EC)
	European Organisation for the Exploitation of Meteorological Satellites (EUMETSAT)
	European Space Agency (ESA)
France	Centre National d'Etudes Spatiales (CNES)
Germany	Deutsche Agentur für Raumfahrt-Angelegenheiten (DARA)
India	Indian Space Research Organisation (ISRO)
Italy	Agenzia Spaziale Italiana (ASI)
Japan	Science and Technology Agency (STA)
	National Space Development Agency (NASDA)
Russia	Russian Federal Service for Hydrometeorology and Environment Monitoring (ROSHYDROMET)
	Russian Space Agency (RSA)
Sweden	Swedish National Space Board (SNSB)
Ukraine	National Space Agency of Ukraine (NSAU)
United Kingdom	British National Space Centre (BNSC)
United States	National Aeronautics and Space Administration (NASA)
	National Oceanic and Atmospheric Administration (NOAA)

CEOS OBSERVERS

Belgium	Federal Office for Scientific, Technical and Cultural Affairs (OSTC)
Canada	Canada Centre for Remote Sensing (CCRS)

New Zealand   Crown Research Institute (CRI)
Norway   Norwegian Space Centre (NSC)

CEOS AFFILIATES

Food and Agriculture Organisation (FAO)
Global Climate Observing System (GCOS)
Global Ocean Observing System (GOOS)
International Council of Scientific Unions (ICSU)
International Geosphere Biosphere Programme (IGBP)
Intergovernmental Oceanographic Commission (IOC)
United Nations Environment Programme (UNEP)
United Nations Office of Outer Space Affairs (UNOOSA)
World Climate Research Programme (WCRP)
World Meteorological Organisation (WMO)

References

Abiodun A A (1993) An international remote sensing system: a possibility, *Space Policy* **9**(3), 179–84.
Ahmad Y J, S El Serafy and E Lutz (1989) *Environmental accounting for sustainable development*, The World Bank, Washington DC.
Anon (1995) *Wall Street Journal* 24 February 1995.
Anon (1996) *Concept for an Integrated Global Observing Strategy*, US Committee on Environment and Natural Resources, Task Force on Observations and Data Management.
Arnaud M (1995) The SPOT programme, *TERRA 2 – Understanding the Terrestrial Environment. Remote Sensing Data Systems and Networks*, ed. P Mather, John Wiley, Chichester, 29–39.
Barrett E C (1974) *Climatology from satellites*, Methuen, London.
Barrett E C and L F Curtis (1982) *Introduction to environmental remote sensing*, second edition, Chapman and Hall, London.
Bartlemus P, C Stahmer and J van Tongeren (1991) Integrated national accounts and economic accounting: framework for a SNA satellite system, *Review of Income and Wealth* **27**(2), 111–48.
Bezy J L, M Rast, S Delwart, P Merhaim-Kealy and S Bruzzi (1996) The ESA Medium Resolution Imaging Spectrometer, *Backscatter* **7**(3), 14–19.
BNSC (1995) *Final report of a working group on the two step approach to the development of a structure for the Earth observation market*, Report to the Earth Observation Programme Board of the British National Space Centre, June 1995.
BNSC (1996) *UK space policy*, British National Space Centre, London.
Boutros-Ghali B (1994) International cooperation in space for security enhancement, *Space Policy* **10**(4), 265–76.
Campbell J B (1996) *Introduction to remote sensing*, Taylor and Francis, London.
CEOS (1994) *Consolidated report*, Committee on Earth Observation Satellites, July 1994.
CEOS (1995) *Coordination for the next decade*, Committee on Earth Observation Satellites, European Space Agency, Paris.
CEOS (1996) *Worldwide directory of on-line services for Earth observation data users*, first edition, CEOS, NASDA, Tokyo.
CES (1995) *Earth observations from space: history, promise and reality*, Committee on Earth Studies, Space Studies Board, National Research Council, Washington DC.
Congressional Record (1996a) *National Defense Authorization Act for Fiscal Year 1997*. Amendment number 4321, p. S6925.
Congressional Record (1996b) *Conference Report on High.R. 3230, National Defense Authorization Act for Fiscal Year 1997*, sec. 1064. Prohibition on collection and

release of detailed satellite imagery relating to Israel.

Cracknell A P and L W B Hayes (1991) *An introduction to remote sensing*, Taylor and Francis, London.

Cudlip W, M Hutchins, S Foley, F Sawdon, C Dorn, P Hicks and J French (1996) *UK Envisat ground segment. User requirements document*, Defence Research Agency, Farnborough, DRA/CIS(CIS2)/5/31/1/2/E24.

Curran P J (1985) *Principles of remote sensing*, Longman, London.

Dreier T (1992) International law perspective on establishing a European regulation with regard to the legal protection of the use of satellite remote sensing data, *Space in the service of the changing Earth*, volume III, eds T D Guyenne and J J Hunt, ESA SP-341, ESTEC, Noordwijk, 1501–4.

Dufresne L (1992) Protection of SPOT data and derived products under private agreements, the law and international conventions, *Space in the service of the changing Earth*, volume III, eds T D Guyenne and J J Hunt, ESA SP-341, ESTEC, Noordwijk, 1471–5.

EOS Payload Advisory Panel (1992) *Adapting the Earth Observing System to the projected $8 billion budget: recommendations from the EOS investigators*, eds B Moore and J Dozier, NASA, Washington DC.

ESA (1989) *Rules concerning information and data*, European Space Agency Council, ESA/C(89)95, rev. 1, Paris, 21 December 1989.

ESA (1991) *Agreement between the European Space Agency and the British National Space Centre concerning the direct reception, archiving, and distribution of ERS-1 data*, Paris, 30 August 1991.

ESA (1994a) *Declaration on the POEM-1 programme. The development and exploitation programme of the First Polar Orbit Earth Observation Mission using the Polar Platform*, drawn up on 27 February 1992, amended on 8 November 1994, ESA/PB-EO/XXVII/Dec 1(final), rev. 4.

ESA (1994b) *Scientific achievements of ERS-1*, ESA SP-1176/I, ESTEC, Noordwijk.

ESA (1994c) *Principles of the provision of ERS data to users*, ESA/PB-EO(90)57, rev. 6, 9 May 1994.

ESA (1994d) *IEOS Data Exchange Principles*, ESA/PB-EO(93)94, annex, European Space Agency, Paris.

ESA (1994e) *Envisat-1 ground segment concept*, ESA/PB-EO(94)24, rev. 3, Paris, 20 September 1994.

ESA (1996a) *Background principles and planning for the establishment of an Envisat data policy*, ESA/PB-EO(96)51, 22 May 1996.

ESA (1996b) *Data operations scientific and technical advisory group, Quarterly statistics on ERS operations*, ESA/PB-EO/DOSTAG(96)4, Paris, 11 January 1996.

ESF (1992) *A strategy for Earth observation from space*, European Science Foundation, Strasbourg, September 1992.

ESYS (1994) *Review of the exploitation of ERS-1 data in Europe and Canada. volume 1: Summary analysis*, ESA/ESRIN ref P340085/PDEX/E, Frascati.

EUMETSAT (1991) *The European Organisation for Meteorological Satellites*, Darmstadt, Germany.

EUMETSAT (1994a) *Background information on the conditions of real time access to EUMETSAT High Resolution Image (HRI) data outside the EUMETSAT member states as agreed by the EUMETSAT Council at its 25th and 26th meetings in June and November 1994*, EUMETSAT, Darmstadt.

EUMETSAT (1994b) *Conditions of real time access to EUMETSAT HRI data outside the EUMETSAT member states*, Resolution EUM/C/94/Res.I, adopted at the 25th meeting of the EUMETSAT Council on 22–24 June 1994, as amended by Resolution EUM/C/94/Res.IV adopted at the 26th meeting of the EUMETSAT Council on 22–24 November 1994.

Eurimage (1996) *Price list*, October 1996.

EUROGI (1996) *Legal protection of geographical information*, EUROGI, The Netherlands.

European Commission (1992a) *Issues in Earth observation data policy for Europe*, Report of a working group, December 1992.

European Commission (1992b) *Issues in Earth observation data policy for Europe*, Final Report by Logica, November 1992, study contract ETES-0018.

European Commission (1996a) Directive 96/9/EC of the European Parliament and of the Council of 11 March 1996 on the legal protection of databases, *Official Journal of the European Communities*, 27 March 1996, L 77/20–28.

European Commission (1996b) *Development and competitiveness of space industries in Europe*, Report of the Industry's High Level Group to the European Commission, European Commission, Brussels.

Forshaw M R B, A Haskell, P F Miller, D J Stanley and J R G Townshend (1983) Spatial resolving power of remotely sensed imagery. A review paper, *International Journal of Remote Sensing* **4**, 497–520.

Fox S, N Prasad and M Szczur (1996) *NASA's Earth Observing System Data and Information System (EOSDIS): an integrated system for processing, archiving, and disseminating high-volume Earth science imagery and associated products*, Hughes Information Technology Systems, Maryland, 215-TP-001-001.

Gates W (1996) *The road ahead*, Penguin, London.

Gaudrat P (1992a) *Aspects juridiques de la teledetection: la protection des donees*, Compte-rendu du 1er workshop, ESA HQ, Paris, 19 June 1992.

Gaudrat P (1992b) *Aspects juridiques de la teledetection: presentation des resultats de l'étude*, second workshop, European Commission, Brussels, 20 October 1992, two volumes.

GCOS (1993) *Draft plan for the Global Climate Observing System (GCOS)*, Joint Scientific and Technical Committee for GCOS, Geneva.

GOOS (1993) *The case for GOOS*, GOOS Publications Series No. 1.

GTOS (1996) *The Global Terrestrial Observing System*, GTOS Secretariat, FAO, Rome.

Harries J (1994) *Earthwatch: the climate from space*, John Wiley, Chichester.

Harris, R (1987) *Satellite remote sensing: an introduction*, Routledge, London.

Harris R (1992) Development issues for the Earth observation ground infrastructure in Europe, *Journal of the British Interplanetary Society* **45**, 111-16.

Harris R (1995) Challenges for the transition of satellite Earth observation to operational status, *EOMARK 95 conference proceedings*, Association Aéronautique et Astronautique de France, Paris, 6.4–6.7.

Harris R and R Krawec (1993a) Some current national and international Earth observation data policies, *Space Policy* **9**(4), 273-85.

Harris R and R Krawec (1993b) Earth observation data pricing policy, *Space Policy* **9**(4), 299-318.

HDP (1994) *Human Dimensions of Global Environmental Change Programme: Work Plan 1994–1995*, HDP.

Houghton J T, L G Meira Filho, B A Callander, N Harris, A Kattenberg and K Maskell (eds) (1995) *Climate change 1995 – The science of climate change: contribution of Working Group I to the second assessment report of the Intergovernmental Panel on Climate Change*, Cambridge University Press, Cambridge.

IACGEC (1992) *Report from IACGEC advisory group 2 on international aspects of global environmental change (GEC) data management*, UK Inter-Agency Committee on Global Environmental Change.

References

IGBP (1992) *Global change: reducing uncertainties*, IGBP, Stockholm.
Jasentuliyana N (1988) United Nations Principles on remote sensing, *Space Policy* 4(4), 281–84.
Jensen W (1995) *How can we make an EO business in Europe*, Presentation to a British Association of Remote Sensing Companies workshop, 2 November 1995.
Krawec R (1995) Ukrainian space policy – contributing to national economic development, *Space Policy* 11(2), 105–14.
Kumon T (1992) Satellite data protection from the point of view of a data user, *Space in the service of the changing Earth*, volume III, eds T D Guyenne and J J Hunt, ESA SP-341, ESTEC, Noordwijk, 1495–500.
Lillesand T M and R W Kiefer (1994) *Remote sensing and image interpretation*, third edition, John Wiley, Chichester.
Logica (1991) *Assessment of remote sensing ground infrastructure requirements*, Final report, Commission of the European Communities, ETES-0004-B.
Logica (1993) *Report on ERS-1 data flows*, UK/National Ground Segment of Polar Platform, Report LSC.302.30510 to DRA, Farnborough.
Logica (1996) *The impact of Envisat operations policy*, Paper 3 of a report on the analysis of the impact of ESA data policy and ESA activities on the implementation of a UK Envisat ground segment, contract CSM/245, DRA/BNSC.
Lutz E and M Munasinghe (1991) Accounting for the environment, *Finance and development*, The World Bank, Washington DC, 19–21.
Macaulay M K (1991) *Testimony before the Committee on Science, Space and Technology*, US House of Representatives, 26 November 1991.
Macaulay M K (1992) *Remotely sensed data from space: distribution, pricing, and applications*, Background paper, Office of Technology Assessment, Congress of the United States, July 1992.
Macaulay M K (1995) NASA's Earth observations commercialization application program, *Space Policy* 1(1), 53–65.
Macaulay M K and M A Toman (1992) Remote sensing of Earth from space, *Resources for the Future* 107, 1–5.
Mansell R, S Paltridge and R Hawkins (1992) *Issues in Earth observation data policy for Europe: industrial dynamics and pricing policies*, Report prepared by the Science Policy Research Unit, University of Sussex for Logica Space and Communications Ltd.
Mansell R and R Hawkins (1992) *Issues in Earth observation data policy for Europe: economic analysis*, Report prepared by the Science Policy Research Unit, University of Sussex for Logica Space and Communications Ltd.
Mather, P (1987) *Computer processing of remotely sensed data – an introduction*, John Wiley, Chichester.
NASA (1996) *The Earth Observer*, 8(5), NASA Goddard Space Flight Center.
NERC (1996) *Natural Environment Research Council data policy handbook*, version 1.0, Swindon.
NOAA (1991) *US Federal Register*, 56(139), 19 July 1991.
NRC (1991) *Solving the global change data puzzle: a US strategy for managing data and information*, National Academy Press, Washington DC.
NSTC (1996) *Our changing planet. The FY 1997 US Global Change Research Program*, Report by the Subcommittee on Global Change Research, Committee on Environment and Natural Resources of the National Science and Technology Council, Washington DC.
Ordnance Survey (1996) *Economic aspects of the collection, dissemination and integration of governments geospatial information*, A report arising from work carried out for

References

Ordnance Survey by Coopers & Lybrand, Ordnance Survey, Southampton.
Pavlakis P, A Sieber and S Alexandry (1996) Monitoring oil-spill pollution in the Mediterranean with ERS SAR, *ESA Earth Observation Quarterly* **52**, 8–11.
Pearce D (1995) *Capturing environmental value. Blueprint 4*, Earthscan, London.
Perry J S (1993) *Understanding our own planet*, ICSU, Paris.
Pryor C, R Harris and J Williams (1996) Looking down on the farm, *IEE Review* **42**(6), 238–42.
Radarsat (1996) *Data license and distribution agreement, commercial appendix*, Radarsat International, Vancouver, Canada.
Shaffer L (1992) US data policy for Earth observations from space, *Space in the service of the changing Earth*, volume III, eds T D Guyenne and J J Hunt, ESA SP-341, ESTEC, Noordwijk, 1477–81.
Shaffer L R and and P Backlund (1990) Towards a coherent remote sensing data policy, *Space Policy* February, 45–52.
Sharman M (1995) Validating the results of the MARS project, *TERRA 2 – Understanding the Terrestrial Environment. Remote Sensing Data Systems and Networks*, ed. P Mather, John Wiley, Chichester, 213–25.
Silvestrini A (1991) *The Landsat program: management, funding and policy decisions*, Submission to the Committee on Science, Space and Technology, US House of Representatives, 26 November 1991.
Smith B (1994) Civil and military space in Europe, *Space Policy* **10**(2), 91–4.
Spero J E (1982) Information: the policy void, *Foreign Policy* **48**, 139–56.
SWCC (1990) *Ministerial declaration*, Second World Climate Conference, 29 October – 7 November, World Meteorological Organisation, Geneva.
Synthesis (1995) *Policy for the dissemination of Earth observation data from space*, Ministere Delegue a la Poste aux Telecommunications at a l'Espace.
Thiebault W (1992) Opening speech chairman of the IISL/ECSL session on legal protection of satellite remote sensing data, *Space in the service of the changing Earth*, volume III, eds T D Guyenne and J J Hunt, ESA SP-341, ESTEC, Noordwijk, 1463–5.
Thiem V (1992) EUMETSAT practice for protection of its satellite data, *Space in the service of the changing Earth*, volume III, eds T D Guyenne and J J Hunt, ESA SP-341, ESTEC, Noordwijk, 1467–70.
Thomas, G B, J P Lester and W Z Sadeh (1995) International cooperation in remote sensing for global change research: political and economic considerations, *Space Policy* **11**(2), 131–41.
Townshend J (1996) Data policy for the IGBP, *IGBP Newlsetter* 27, September 1996.
Triebnig, G (1994) The ESA Earth-Observation Guide and Directory Service, *ESA Bulletin* **78**.
Triebnig G (1995) The ESA Earth observation Guide and Directory Service, *TERRA 2 – Understanding the Terrestrial Environment. Remote Sensing Data Systems and Networks*, ed. P Mather, John Wiley, Chichester, 129–39.
United States (1990) *Assessment of fees for access to environmental data*, US Law §1534.
Williams J (1995) *Geographic information from space*, John Wiley, Chichester.

Index

absorption 10, 12
access 45, 50, 51, 58, 83, 85, 88, 89, 97, 109, 128, 132
accounted costs 125
ADEOS 9, 17, 22, 51, 68
Advanced Along Track Scanning Radiometer 135
Aerospatiale 136
agreements 91, 99
agriculture 100, 122, 138
airborne data 8, 18
ALMAZ 18
Along Track Scanning Radiometer 9, 12, 16, 34, 59, 70, 77, 79
ALOS 17, 26
amplitude 8
Announcement of Opportunity 52, 89, 90, 109, 117
applications 41, 53, 57, 84
archives 5, 48, 50, 54, 55, 64, 77, 79, 82, 85, 126, 132, 140, 143
Argentina 26, 27, 45, 68, 139
Artemis 70, 71
ASAR 80, 82, 118, 132, 135
Aswan High Dam 128
atmospheric chemistry 12, 16, 19, 30
audit trail 130
Australia 135, 139
AVHRR 11, 34, 67, 70, 113, 115
AVNIR 9

Ball Aerospace 19
bandwidth 129
barriers 106, 111, 114, 135
benefits 21, 23, 38, 44, 46, 47, 48, 84, 90, 103, 104, 108, 117, 121
blackbody radiation 9, 10

Boltzman's constant 10
Brazil 15, 139
Brest 79
British National Space Centre 2, 63, 86, 118, 126, 137
browse 131
buyer 100, 114, 120, 121, 123, 125

calibration 49, 89, 130, 131
Canada 56
Canada Centre for Remote Sensing 15
Canadian Space Agency 1, 2, 16
carbon dioxide 10
catalogue 77, 84, 85, 131
categories 58, 96, 98, 103, 108, 109, 124, 131, 141
CBERS 16
Centre for Earth Observation 15, 83, 129, 143
China 15, 16
climate 3, 18, 23, 32, 33, 34, 59, 63, 128
closed loop 105, 107, 112, 120
CNES 94
Columbus 126
commercial 6, 24, 39, 41, 42, 57, 60, 61, 62, 77, 86, 87, 90, 93, 96, 97, 103, 106, 113, 115, 118, 121, 137, 140
Committee on Earth Observation Satellites 15, 26, 27, 28, 35, 43, 46, 47, 48, 49, 50, 51, 54, 55, 59, 65, 87, 142, 144
communications 129, 132
conflict 25, 66, 134
Consultative Committee on Space Data Systems 88
continuity 24, 34, 60, 62, 64, 141
contract 92
copyright 92, 94

Index

cost 45, 50, 52, 53, 60, 61, 64, 67, 68, 73, 75, 85, 89, 90, 91, 93, 96, 104, 106, 115, 132
cost recovery 103, 104, 105
creativity 92
criteria 51, 52, 55, 64, 132

data acquisition 51, 72
data capture 132
data chain 67, 114
data conservation 127
data distribution 1, 50, 60, 77, 82, 85, 86, 88, 89, 95
data exchange 48, 49, 50, 52, 53, 55, 63, 64, 106
data grants 125
data granule 131
data management 34, 63, 64, 82, 130, 132
data processing 51
data protection 5, 91, 94
data quality 48, 49, 64, 130
data reception 67, 68, 77, 83, 97
data relay 70, 71, 81, 83
data selection 51
data storage 88, 128
data type 132
data volume 27, 48, 83
database 94, 95, 129, 139
decisions 36, 37
decryption 97, 98
Department of the Environment 135
Department of Trade and Industry 22
Detailed Mission Operations Plan 72
direct broadcast 85
directory 131
disasters 73
discrimination 45, 50, 51, 53, 96
Distributed Active Archive Centres 83, 129
distributors 58, 72, 77, 86
diversity 138
downlink 13, 70, 75, 77, 85

E-Systems 75
Earth Explorer 6
Earth Observer 16, 119
Earth Observing System 6, 15, 19, 25, 26, 51, 64, 70, 82, 85, 139
Earth Observing System Data and Information System 65, 82, 84, 129

EarthWatch 19, 39
economic characteristics 104
economic welfare 102
electromagnetic radiation 8, 93
electronic information 35, 36, 95, 98
emissivity 10
enabling services 84
encryption 36, 61, 97, 98, 139
environment 2, 26, 27, 32, 39, 44, 48, 62, 104, 127, 141
environmental accounting 39
environmental agreements 29, 38, 39
environmental change 63
environmental value 38
Envisat 3, 13, 16, 26, 27, 39, 51, 59, 63, 70, 71, 80, 82, 85, 89, 118, 132, 135, 138, 139, 140, 141
EOSAT 1, 40
ERS 9, 12, 13, 16, 26, 34, 37, 39, 41, 50, 57, 68, 72, 77, 78, 86, 91, 117, 118, 127, 140, 141
ERS Consortium 58, 72, 77
ESOC 72, 77
ESRIN 72, 77, 81
EUMETSAT 3, 6, 14, 16, 23, 24, 43, 56, 58, 60, 62, 91, 93, 97, 98, 106, 112, 115, 119, 142
EUMETSTAT Polar System 16, 60
Eurimage 58
Europe 23, 25, 39, 51, 56, 68, 81, 94, 95, 96, 109, 115, 138
European Association of Remote Sensing Companies 138
European Centre for Medium Range Weather Forecasting 58
European Commission 2, 15, 23, 51, 84, 94, 98, 105, 138, 141, 143
European Parliament 94
European Science Foundation 43, 56, 61, 66, 125
European Space Agency 2, 3, 15, 16, 22, 23, 26, 28, 37, 40, 42, 43, 50, 51, 53, 56, 57, 62, 71, 74, 80, 83, 90, 105, 107, 112, 117, 119, 127, 128, 135
European Wide Service Exchange 84, 129
exclusive use 50, 51
experimental programmes 63

fair price 120, 121
fair treatment 7

Farnborough 79
finance sector 137
fixation 92
forces 101
foreign station 77, 79, 86
formats 88
France 62, 93, 102, 107
free data 111
frequency 9
Fucino 79
full price 117
funding structures 106

Gatineau 79
GCOS 32, 110
geographical information 92
geographical territory 58, 86
geospatial 36, 102, 104
geostationary 3, 13, 14, 16, 17, 18, 60, 67, 70
global change 6, 22, 36, 48, 63, 64, 132
global change research 3, 4, 15, 17, 22, 25, 47, 48, 55, 64, 82, 103, 114, 127, 130, 135
Global Change Research Act 64
Global Change Research Program 22, 64, 135
GMS 17, 67, 85
GOES 13, 18, 67, 85
GOMS 18
GOOS 32, 132
government customer 135
government policy 20, 21, 36, 47, 63, 103, 105, 124, 135
government sponsorship 22, 25, 135
ground segment 67, 75, 76, 77, 78, 80, 82, 85, 121, 130, 139
ground station 67, 68, 70, 74, 75, 95
growth 7, 26, 27, 28, 134
GTOS 22
guide 131

hazards 71
HDP 30
Hughes 136

IEOS 51, 55, 65, 109
IGBP 3, 27, 29, 30, 43, 51, 54, 66, 110
IGOS 34, 35
India 26, 139
Indian Space Research Organisation 17

industry 56, 136
information content price 121, 139
information products 36, 39, 80, 81, 83, 87, 122
information vouchers 125, 142
INSAT 17
instrument 14, 23, 26, 73, 80, 82, 89, 93
intellectual property rights 2, 52, 56, 90, 92
Intelligent Satellite data Information System 131
interface 129
Intergovernmental Panel on Climate Change 32
International Space Year 126
Internet 84, 85, 88, 104, 129, 140
interrogation 131
inventory 131
IRS 17
Israel 46

Japan 17, 22, 26, 51, 138
Japanese space agency 15
JERS-1 9, 17, 26

Kiruna 68, 70, 75, 79, 81

land cover/use 30, 33, 44
Land Remote Sensing Policy Act 127
Landsat 1, 11, 12, 13, 14, 19, 23, 26, 34, 39, 40, 41, 49, 68, 70, 104, 116, 117, 127, 128, 131, 140
laser 26
legal 47, 58, 91
legislation 95
licence 1, 2, 36, 58, 68, 75, 86, 98, 99
lidar 15
line of sight 67, 68, 70
Lockheed Martin 19, 75, 136
Logica 73
Low rate Reference Archive Centre 81

marginal cost price 110, 113
marine biology 80
market structure 137
markets 24, 37, 39, 40, 96, 100, 104, 105, 114, 119, 123, 124, 136, 138
MARS programme 73
Maspalomas 79
Matera 79
Matra 68, 136

maximising use   7, 19, 27, 49, 103, 105, 106, 108, 143
MERIS   12, 80, 82
metadata   34, 48, 49, 55, 63, 64, 77, 84, 85, 130
METEOR   18
meteorology   3, 6, 16, 18, 60, 77, 97, 103, 108
Meteosat   9, 11, 12, 16, 23, 24, 60, 67, 70, 85, 94, 97, 98, 112, 139
Metop   16, 59, 139
microwave   9, 16, 17, 26
middle infrared   9
Mission to Planet Earth   6, 19, 64, 65, 82, 135, 143
Mitsubishi   75
mobile station   68
monitoring   16, 27, 32, 36, 47, 48, 60, 63, 139
monopoly   96, 105, 120
MOS   17

NASA   15, 17, 18, 19, 37, 40, 43, 50, 51, 64, 82, 105, 108, 116, 120, 128, 135, 142
NASDA   37, 51, 83, 105
National Remote Sensing Centre Limited   136
National Research Council   130
National Satellite Land Remote Sensing Data Archive   127, 132
national security   75
national station   77, 79, 86
near infrared   9, 11, 17
NERC   63, 138
networks   75, 82, 88, 140, 143
Nile river   128
NOAA   11, 13, 14, 15, 18, 19, 24, 40, 51, 67, 70, 113, 116, 117, 120
non-discriminatory   45, 50, 51, 53

Oberpfaffenhofen   79
objectives   102, 105,125, 139, 140, 141
oceans   17, 19, 33, 38, 79, 132
Odin   18
off-line   77, 81, 82
oil slicks   101, 121
Okean   18
open loop   105, 107, 112
operational   6, 24, 32, 51, 60, 61, 63, 87, 106, 113
optical   17, 19

Orbimage   19, 46
originality   93
ownership   58, 60, 68, 89, 90, 135
oxygen   10
ozone   10, 15, 16, 38, 59

Payload Data Acquisition Station   81
Payload Data Handling Station   81
Payload Data Segment   81
pedigree   130
phase   8
pilot project   117
pixel   12, 17, 74
Planck's radiation law   9
plane waves   8
pointing   73, 75
Polar Platform   19, 51
policing   120, 121
Potato Marketing Board   73
potato monitoring   73, 74
preferential treatment   24, 108, 123
Preferred Exploitation Plan   72
preservation   48, 50
price   48, 50, 58, 59, 96, 100, 138, 141
Primary Data User Station   97, 98
principal investigator   50, 55, 65, 107, 117, 125
principles   44, 47, 51, 52, 54, 55, 56, 62, 63
private sector   56, 57, 91, 102, 135, 136, 138
Processing and Archiving Centre   81
Processing and Archiving Facility   77, 79, 81
product level   132
products   80, 81, 83, 87, 98, 99, 123, 138
programming   72, 73, 93
protection   121
public benefit   6, 22, 47, 48, 50, 51, 52, 53, 60, 102, 104, 105, 108, 119, 123
public good   21, 22, 26, 36, 54, 109, 123, 124, 134
public law   120, 121
public policy   2, 21, 37, 38, 134
public sector   37, 57, 91, 102, 134, 137, 143
publication   52, 89, 96, 109

radar   9, 12, 17, 19, 59, 72, 141
Radarsat   1, 2, 16, 27, 58, 119
radiometer   14, 16
RAIDS   77

Index

realisable price 114
rebalancing 100, 123
receiving station 67, 68, 69, 70, 71
research use 53, 87, 96, 104, 109, 112, 115, 119, 121, 123
resolution 12, 13, 17, 19, 29, 46
Resource satellite 18
resources 2, 3, 36, 94, 123, 126, 142
return on investment 60, 62, 91, 103, 139
revisit 13
risk 63
Russia 18
Rutherford Appleton Laboratory 77

SAR 9, 13, 14, 25, 59, 68, 70, 72, 77, 79, 86, 87, 90, 117, 118, 121, 132
scene 121
scientific research 6, 7, 23, 26, 27, 29, 56, 57, 59, 61, 95, 106, 139
Sea Empress 73
search engines 129
Seastar 19
Secondary Data User Station 97, 98
seller 114, 120
Shuttle 19, 25
Shuttle Imaging Radar 9
SICH 18
simplicity 7
software 84, 99, 139
Space Imaging 19, 27, 39, 73, 74, 102
space segment 85, 105, 130
Space Station 126
SPOT 9, 12, 13, 14, 15, 17, 18, 23, 26, 34, 39, 49, 58, 62, 68, 69, 73, 74, 75, 76, 77, 85, 86, 93, 95, 102, 105, 107, 127, 130, 131
stability 6, 24, 139
standards 49, 55, 63, 64, 87
Stefan–Boltzmann constant 10
strategy 60, 84, 130, 138
substitute 100
sui generis 94, 95
sun synchronous 13
supplier 100, 112, 119, 139
sustainability 20, 21, 22, 118, 135
sustainable development 2, 3, 7, 27, 33, 142
system engineering 136

tape recorders 70, 75, 93

tapes 128
TDRSS 70
technology 8, 40, 48, 49, 56, 57, 87, 88, 128, 135
temperature 10, 14
Thailand 68
Thematic Mapper 11, 12, 70, 104
thermal infrared 9, 11, 17
TOMS 19
TOPEX/Poseidon 17, 64
Toulouse 73, 75, 77, 85, 93
transition 113, 117, 118, 119, 125, 138
transparency 139
trends 88, 98, 139
TRMM 17, 22, 26, 51, 64
two-step approach 137
two-tier price 119

UARS 19, 23
UK 63, 77, 89, 102
Ukraine 18
UNESCO 126
United Nations 45, 46
United States 18, 22, 25, 39, 40, 46, 51, 64, 68, 87, 92, 93, 98, 106, 109, 111, 120, 138, 143
US Senate 46
use 47, 48, 49, 52, 84, 96, 141
users 6, 21, 23, 27, 50, 51, 52, 54, 60, 65, 72, 97, 100, 105, 109, 115, 123, 125, 132, 137, 138, 139

validation 49, 50, 51
value 20, 21, 38, 48, 60, 64, 100, 106, 107, 109, 112, 119, 120, 121, 123, 134, 139, 143
value added companies 136
value chain 24, 37, 60
visible wavelengths 9, 11, 17
visualisation 131

water vapour 10, 12
watermark 99
wavelength 8, 9, 10, 14, 17
waves 8
WCRP 29
weather satellites 3
West Freugh 77, 86, 87
Wien's displacement law 9
World Wide Web 15, 77, 83, 84, 131
WorldMap 18